The linearized theory of gravitational radiation, and the detection of gravitational waves.

By Petros Souvatzis

Preface

The idea of writing this thesis was born simultaneously with the construction of The Laser Interferometer Gravitational-Wave Observatories (LIGO), placed in Livingston, Louisiana and Hanford, Washington (both in the United States). The idea to publish this thesis in the form of a book was inspired by the announcement of the first successful detection of gravitational waves February 11th 2016. This book can be seen as a homage to all the brave scientist that made this detection possible.

The fascinating thing about the LIGO-observatories is that they must be able to measure gravitationally wave induced length differences which are 1/1000 the diameter of an atom, in order to detect gravitational waves. In this paper a conceptual explanation of how these length differences can be measured, and a pedagogic review of the linearized gravitational wave theory will be presented.

The most useful result presented in this thesis, is that the linearized gravitational wave theory, almost alone, manages to come up with nearly the same predictions of which gravitational wave sources that can be expected to be detected by LIGO, as the more advanced studies made by the scientists working with LIGO. The main difference between the considerations made in this thesis and the more advanced ones, is that the more advanced studies, in addition to the linearized theory have used estimates of the LIGO-interferometers signal to noise ratios, whereas the studies presented here almost exclusively use the linearized gravitational wave theory.

The estimates made in this thesis, in accord with the previously mentioned more exact estimates, predict that gravitational waves emitted from coalescing black-hole binaries, at distances 200 Mpc and 700 Mpc, together with the waves emitted by coalescing Neutron-star binaries, at distances 23 Mpc, 60 Mpc and 200 Mpc, will be detectable by LIGO.

Finally, I whish to express my profound gratitude to Professor Ulf Danielsson for his support during the writing of this book.

Petros Souvatzis

Contents

Introduction

Ever since Joseph Weber in the 1960's built his first resonance detector in an attempt to detect gravitational waves, physicists have been concerned with the theory and the detection of gravitational waves. But up to this date no one has actually succeeded in detecting these remarkable waves. Thus, even though the theory of general relativity predicts the existence of these waves, it still remains to be seen if gravitational waves really exist. So the search of gravitational waves is not only a quest for the detection of a new physical phenomenon, it also represents one of the most challenging tests of the theory of general relativity.

Despite the fact that gravitational waves have never been detected, their existence can be made plausible by indirect observations. Such observations have been made by the two American physicists Hulse and Taylor. They discovered that the rate at which the neutron star binary PSR1916+13 decreased its period time, was almost the same as the decrease in period time predicted by general relativity. Since the prediction made by general relativity is based on the assumption that the binary loses potential energy due to the emission of gravitational waves, the discovery made by Hulse and Taylor is considered to be the strongest evidence of the existence of gravitational waves.

But what are gravitational waves, and how are they generated? In analogy with Maxwell's electromagnetic field equations that predict the emission of electromagnetic waves whenever charged particles are being accelerated, the field equations of general relativity (Einstein's equations) predict the emission of gravitational waves, whenever massive bodies are being accelerated. These gravitational waves can be thought of as, periodic disturbances in the gravitational field, or maybe more accurately, as periodic changes in the space-time curvature, travelling at the speed of light.

The theory of general relativity also predicts that whenever a gravitational wave passes through an object it will stretch the object's dimensions in one direction, while shrinking it in a direction orthogonal to the stretching direction. It is by utilizing these stretching and shrinking effects that one hopes to detect gravitational waves.

The oldest method used in trying to detect gravitational waves uses resonance bars. These detectors are constructed of massive aluminium cylinders that can be thought of as huge harmonic oscillators. When a gravitational wave reaches the detector, the stretching and shrinking effects of the wave will force the bar into oscillation. Whenever a gravitational wave with the proper frequency "hits" the bar, the detector will be excited into resonance, enabling the detection of the wave as an excitation of the bar's energy state. Unfortunately the resonance detectors have not been accurate enough in providing direct evidence to the existence of gravitational waves.

The setback in the use of resonance detectors has motivated the scientists to develop another kind of gravitational wave detectors. These detectors are huge state of the art Michelson-interferometers, with interferometer arm lengths ranging from 300m up to 4 km. The arms of these interferometers are placed at right angles relative to each other, in a L-like shape (see figure 0 at next page). At the end and the beginning of each arm, mirrors are suspended with the freedom to move back and forth along the directions of the interferometer arms.

When a gravitational wave "hits" the interferometer, the distance between the mirrors in one of the arms will decrease whereas the distance between the mirrors in the other arm will increase (see figure below).

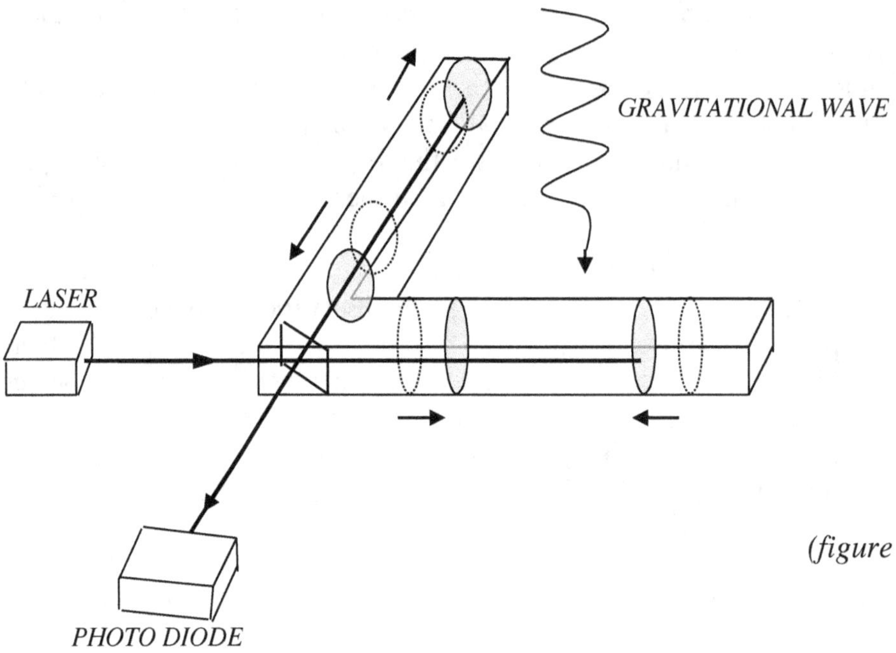

(figure 0)

By splitting the laser beam, so that equal parts of the beam enter each interferometer arm, and by realigning the two beams on a photo-diode, after they have been reflected back and forth several times, the gravitational wave can be detected by analysing the "interference-pattern" projected on the diode.

Several interferometer-based gravitational wave detectors are now being under construction. The construction of these detectors is a slow and demanding process, since these detectors must be able to detect phase-shifts, corresponding to length differences equal to 1/1000 the diameter of the nucleus of an atom. The detectors are estimated to be completed 2002.

The goals of this thesis are:

- To present a pedagogical review of the gravitational wave theory, based on Einstein's linearized equations of general relativity.
- Explain at an undergraduate level how the gravitational wave detectors used by LIGO work (LIGO stands for the Laser Interferometer Gravitational-Wave Observatories, placed in Livingston Louisiana and Hanford, Washington; Both in the United States).
- With the help of the linearized gravitational wave theory and the predicted sensitivities of the LIGO interferometers, estimate which gravitational wave sources that can be expected to be detected.

1 Differential geometry

In this chapter some basic concepts of differential geometry will be introduced. This is done because the theory of gravitational radiation has mainly been developed within the context of general relativity, and since differential geometry is one of the most important mathematical tools used in the study of general relativity.

This chapter requires that the reader has some previous knowledge of two-dimensional curved surfaces that are imbedded in the three-dimensional Euclidean space, and tensors.

The goal of this chapter is not to present the complete theory of differential geometry, but rather to give the reader the ideas that are necessary for the understanding of the theory of gravitational radiation.

1.1 Cartesian coordinates

Since the derivative and the parallel transport of a vector are the most fundamental concepts used in any physical theory, this chapter will almost entirely be devoted to discussing these fundamental concepts. We will begin by studying derivatives of vectors and tensors in Euclidian space. This will be done in two specific cases. In the first case the vectors will be expressed in Cartesian coordinates, and in the second case they will be expressed in curvilinear coordinates. This is done so that we may develop the mathematical tools necessary to the final stages of this chapter. There, we will discuss how the derivatives of vectors and tensors together with the parallel transport of a vector are defined in a curved space. But first let us begin our study in the Euclidean space.

In the n-dimensional Euclidean space of E^n, the position of a point in Cartesian coordinates

is described by the vector

$$\bar{v} = \sum_{\alpha=1}^{n} \hat{x}_{\alpha} v^{\alpha} = \hat{x}_{\alpha} v^{\alpha} \qquad \textbf{(1.1.1)}$$

Where v^{α} is the components of the vector and $\hat{x}_{\alpha}$ is the orthonormal Cartesian basis vectors.

The metric of E^n in Cartesian coordinates is given by $g_{\alpha\beta} = \hat{x}_{\alpha} \cdot \hat{x}_{\beta} = \delta_{\alpha\beta}$ **(1.1.2)**

And finally the derivative of a vector in E^n, with respect to the coordinate x^{β}, is given by:

$$\frac{\partial \bar{v}}{\partial x^{\beta}} = \frac{\partial v^{\alpha}}{\partial x^{\beta}} \hat{x}_{\alpha} \qquad \textbf{(1.1.3)}$$

Observe that in the above discussion the Einstein convention for summing has been used. This convention simply states that whenever two indexes are denoted with the same letter, a summation over these indexes are implicated. This convention will be used throughout this thesis.

1.2 Curvilinear coordinates

Sometimes the geometry of the physical reality that we try to describe has a certain symmetry. The description of this reality is often simplified if one, instead of using Cartesian coordinates, uses a set of coordinates that possess the same symmetry. These coordinates are called curvilinear coordinates, and below two examples of such coordinates have been given.

Spherical coordinates (r, θ, ϕ)

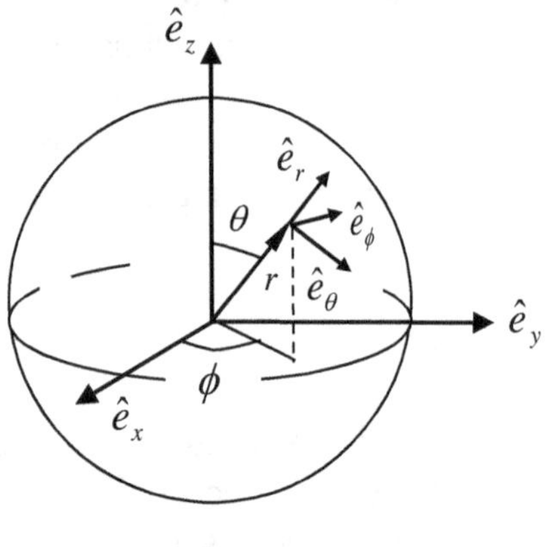

Basis vectors

$$\hat{e}_r = \sin(\theta)\cos(\phi)\hat{e}_x + \sin(\theta)\sin(\phi)\hat{e}_y + \cos(\theta)\hat{e}_z$$
$$\hat{e}_\theta = \cos(\theta)\cos(\phi)\hat{e}_x + \cos(\theta)\sin(\phi)\hat{e}_y - \sin(\theta)\hat{e}_z$$
$$\hat{e}_\varphi = -\sin(\theta)\sin(\phi)\hat{e}_x + \sin(\theta)\cos(\phi)\hat{e}_y$$

Cylindrical coordinates (z, ρ, ϕ)

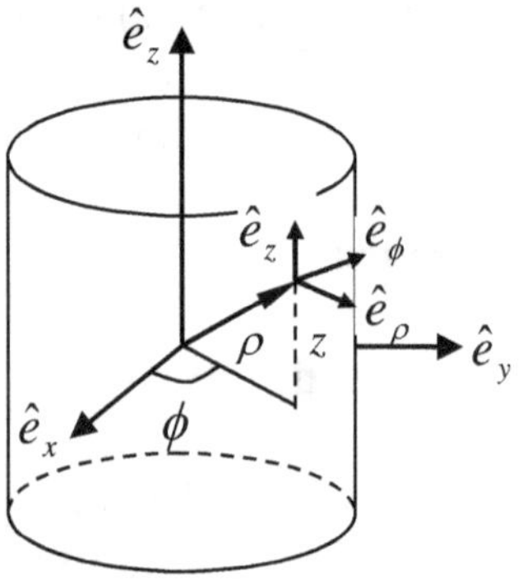

Basis vectors

$$\hat{e}_z$$
$$\hat{e}_\rho = \cos(\phi)\hat{e}_x + \sin(\phi)\hat{e}_y$$
$$\hat{e}_\phi = -\sin(\phi)\hat{e}_x + \cos(\phi)\hat{e}_y$$

The crucial difference between Cartesian coordinates and Curvilinear coordinates in Euclidean space is that the basis vectors of Cartesian coordinates do not share their independence of the coordinates with the Curvilinear coordinates. This means that the Curvilinear coordinates may change in both length and direction. This has two interesting consequences that take the difference one step further:

a) The metric in E^n expressed in Curvilinear coordinates can not be independent of the coordinates, since the metric is expressed by $$g_{\alpha\beta} = \vec{e}_\alpha \cdot \vec{e}_\beta \qquad \textbf{(1.2.1)}$$ where $\vec{e}_\alpha$ is the basis vectors of the Curvilinear coordinates.

b) The derivative of a vector $\vec{v} = \vec{e}_\alpha v^\alpha$ in E^n, with respect to the coordinate y^β, is given by:

$$\frac{\partial \vec{v}}{\partial y^\beta} = \frac{\partial v^\alpha}{\partial y^\beta}\vec{e}_\alpha + v^\alpha \frac{\partial \vec{e}_\alpha}{\partial y^\beta} = v^\alpha{}_{,\beta}\vec{e}_\alpha + v^\alpha \frac{\partial \vec{e}_\alpha}{\partial y^\beta} \quad ; \text{ where } \quad v^\alpha{}_{,\beta} \equiv \frac{\partial v^\alpha}{\partial y^\beta} \qquad \textbf{(1.2.2)}$$

1.3 Dual basis vectors and dual vector components

The dual basis vectors denoted $\vec{e}^{\alpha}$ are defined by $\vec{e}^{\alpha} \equiv g^{\alpha\beta}\vec{e}_{\beta}$ **(1.3.1)**

where $g^{\alpha\beta}$ are the inverse components of the metric matrix $g_{\alpha\beta}$.

This in turn implies that $\vec{e}^{\alpha}\vec{e}_{\beta} = g^{\alpha\sigma}\vec{e}_{\sigma}\vec{e}_{\beta} = g^{\alpha\sigma}g_{\sigma\beta} = \delta^{\alpha}_{\beta}$ **(1.3.2)**

The dual vector components v_{α} to a vector $\vec{v} = v^{\alpha}\vec{e}_{\alpha}$ are defined by $v_{\alpha} \equiv g_{\alpha\beta}v^{\beta}$ **(1.3.3)**

1.4 The Christoffel symbols

Since the basis vectors $\vec{e}_{\alpha}$ of the curvilinear coordinates span the entire space E^{n}, the derivative with respect to the coordinate y^{β} of a curvilinear basis vector can be expressed as a linear combination of the same basis vectors, i.e. $\dfrac{\partial \vec{e}_{\alpha}}{\partial y^{\beta}} = \vec{e}_{\mu}\Gamma^{\mu}_{\alpha\beta}$ **(1.4.1)**

Where the linear coefficients $\Gamma^{\mu}_{\alpha\beta}$ are continuous functions, called the Christoffel symbols. Similarly, the derivative of the dual basis vectors $\vec{e}^{\alpha}$ with respect to the coordinate y^{β} can together with the same Christoffel symbols, and the fact that $\vec{e}^{\alpha} \cdot \vec{e}_{\beta} = \delta^{\alpha}_{\beta}$, be defined as

$$\frac{\partial \vec{e}^{\alpha}}{\partial y^{\beta}} = -\vec{e}^{\mu}\Gamma^{\alpha}_{\mu\beta} \qquad \textbf{(1.4.2)}$$

1.5 The covariant derivative

Since the basis vectors in curvilinear coordinates are dependent on the coordinates themselves, we have to redefine the concept of a derivative. This new definition will take into account this dependence and it will be referred to as the covariant derivative.

a) The covariant derivative of a vector component v^{α} with respect to the coordinate y^{β} is defined by

$$\nabla_{\beta}v^{\alpha} = v^{\alpha}{}_{;\beta} \equiv v^{\alpha}{}_{,\beta} + v^{\mu}\Gamma^{\alpha}_{\mu\beta} \qquad \textbf{(1.5.1)}$$

Then by applying (1.3.1) and (1.4.1) to equation (1.2.2), one gets

$$\frac{\partial \vec{v}}{\partial y^{\beta}} = v^{\alpha}{}_{;\beta}\vec{e}_{\alpha} \qquad \textbf{(1.5.2)}$$

b) The covariant derivative of a covector v^{α} with respect to the coordinate y^{β} is then given in a similar way by

$$\nabla_{\beta}v_{\alpha} = v_{\alpha;\beta} \equiv v_{\alpha,\beta} - v_{\mu}\Gamma^{\mu}_{\alpha\beta} \qquad \textbf{(1.5.3)}$$

c) The covariant derivative of a second order covariant tensor $\overline{\overline{T}}$ with respect to the coordinate y^{μ} is defined by

$$\nabla_{\mu}T_{\alpha\beta} = T_{\alpha\beta;\mu} \equiv \frac{\partial \overline{\overline{T}}}{\partial y^{\mu}}(\vec{e}_{\alpha}, \vec{e}_{\beta}) \qquad \textbf{(1.5.4)}$$

Since this definition does not give any direct way of how to calculate the covariant derivative of the tensor mentioned, a more explicit definition can be made by the following consideration

$$\frac{\partial T_{\alpha\beta}}{\partial y^{\mu}} = T_{\alpha\beta,\mu} = \frac{\partial \overline{\overline{T}}}{\partial y^{\mu}}(\vec{e}_{\alpha},\vec{e}_{\beta}) + \overline{\overline{T}}(\frac{\partial \vec{e}_{\alpha}}{\partial y^{\mu}},\vec{e}_{\beta}) + \overline{\overline{T}}(\vec{e}_{\alpha},\frac{\partial \vec{e}_{\beta}}{\partial y^{\mu}}) \Rightarrow$$

$$T_{\alpha\beta,\mu} = \frac{\partial \overline{\overline{T}}}{\partial y^{\mu}}(\vec{e}_{\alpha},\vec{e}_{\beta}) + \overline{\overline{T}}(\vec{e}_{\sigma}\Gamma^{\sigma}_{\alpha\mu},\vec{e}_{\beta}) + \overline{\overline{T}}(\vec{e}_{\alpha},\vec{e}_{\sigma}\Gamma^{\sigma}_{\beta\mu}) =$$

$$= T_{\alpha\beta;\mu} + \Gamma^{\sigma}_{\alpha\mu}T_{\sigma\beta} + \Gamma^{\sigma}_{\beta\mu}T_{\alpha\sigma} \quad \Rightarrow$$

$$T_{\alpha\beta;\mu} = T_{\alpha\beta,\mu} - \Gamma^{\sigma}_{\alpha\mu}T_{\sigma\beta} - \Gamma^{\sigma}_{\beta\mu}T_{\alpha\sigma} \qquad \textbf{(1.5.5)}$$

This is a more useful definition of the covariant derivative of a second order covariant tensor. In almost the same manner it is then easy to derive the expressions for the covariant derivative of a second order contravariant tensor and a first order covariant and a first order contravariant tensor:

$$T^{\alpha\beta}{}_{;\mu} = T^{\alpha\beta}{}_{,\mu} + \Gamma^{\alpha}_{\sigma\mu}T^{\sigma\beta} + \Gamma^{\beta}_{\sigma\mu}T^{\alpha\sigma} \qquad \textbf{(1.5.6)}$$

$$T^{\beta}_{\alpha;\mu} = T^{\beta}_{\alpha,\nu} + \Gamma^{\beta}_{\sigma\mu}T^{\sigma}_{\alpha} - \Gamma^{\sigma}_{\alpha\mu}T^{\beta}_{\sigma} \qquad \textbf{(1.5.7)}$$

It is now simple to calculate the covariant derivative of the metric tensor $g_{\alpha\beta}$:

$$g_{\mu\nu;\sigma} = g_{\mu\nu,\sigma} - \Gamma^{\alpha}_{\mu\sigma}g_{\alpha\nu} - \Gamma^{\alpha}_{\nu\sigma}g_{\mu\alpha} \qquad \textbf{(1.5.8)}$$

But

$$\frac{\partial g_{\mu\nu}}{\partial y^{\sigma}} = \frac{\partial(\vec{e}_{\mu}\cdot\vec{e}_{\nu})}{\partial y^{\sigma}} = \frac{\partial \vec{e}_{\mu}}{\partial y^{\sigma}}\cdot\vec{e}_{\nu} + \vec{e}_{\mu}\cdot\frac{\partial \vec{e}_{\nu}}{\partial y^{\sigma}} = \vec{e}_{\nu}\cdot\vec{e}_{\alpha}\Gamma^{\alpha}_{\mu\sigma} + \vec{e}_{\mu}\cdot\vec{e}_{\alpha}\Gamma^{\alpha}_{\nu\sigma} = g_{\nu\alpha}\Gamma^{\alpha}_{\mu\sigma} + g_{\mu\alpha}\Gamma^{\alpha}_{\nu\sigma} \qquad \textbf{(1.5.9)}$$

Which means that equation (1.5.8) and (1.5.9) together imply:

$$g_{\mu\nu;\sigma} = 0 \qquad \textbf{(1.5.10)}$$

This reveals that the metric tensor (and its inverse) can be used to higher and lower indexes inside covariant derivatives, i.e.

$$g_{\alpha\beta}v^{\beta}{}_{;\mu} = (g_{\alpha\beta}v^{\beta})_{;\mu} = v_{\alpha;\mu} \qquad \textbf{(1.5.11)}$$

In this section, we have discussed the concept of the covariant derivative in order to understand how one can calculate derivatives of vectors and tensors expressed in curvilinear coordinates. But the use of the covariant derivative in connection with curvilinear coordinates is not the only application of this derivative. In fact, the use of the covariant derivative will later turn out to be the simplest way of calculating derivatives of vectors and tensors defined within a curved space.

1.6 Calculation of the Christoffel symbols with the help of the metric

Since the Christoffel-symbols are crucial to the calculations of covariant derivatives, we must find a way of calculating these symbols. In this section we will show that the Christoffel symbols can be expressed only in terms of the metric. We will begin by showing that the Christoffel-symbols are symmetric, i.e. $\Gamma^{\sigma}_{\alpha\beta} = \Gamma^{\sigma}_{\beta\alpha}$.

First let us consider a scalar field ϕ, where the following is true in Cartesian coordinates:

$$\phi_{,\alpha;\beta} = \phi_{,\alpha,\beta} \qquad \textbf{(1.6.1)}$$

since $\Gamma^{\sigma}_{\alpha\beta} = 0$ in Cartesian coordinates. This implies then together with

$$\frac{\partial^2 \phi}{\partial x^{\alpha} \partial x^{\beta}} = \frac{\partial^2 \phi}{\partial x^{\beta} \partial x^{\alpha}} \qquad \textbf{(1.6.2)}$$

that

$$\phi_{,\alpha;\beta} = \phi_{,\beta;\alpha} \qquad \textbf{(1.6.3)}$$

in Cartesian coordinates. But since equation (1.6.3) is a tensorial relation it must be true in all bases. The next step is to express $\phi_{\alpha;\beta}$ with the help of the Christoffel symbols:

$$\phi_{,\alpha;\beta} = \phi_{,\alpha\beta} - \phi_{,\sigma}\Gamma^{\sigma}_{\alpha\beta} \qquad \textbf{(1.6.4)}$$

Then (1.6.3) and (1.6.4) imply that:

$$\phi_{,\alpha\beta} - \phi_{,\sigma}\Gamma^{\sigma}_{\alpha\beta} = \phi_{,\beta\alpha} - \phi_{,\sigma}\Gamma^{\sigma}_{\beta\alpha} \quad \Rightarrow$$

$$\Gamma^{\sigma}_{\alpha\beta} = \Gamma^{\sigma}_{\beta\alpha} \qquad \textbf{(1.6.5)}$$

We are now ready to derive the Christoffel symbols from the metric. The first step towards our goal is to make use of (1.5.9) and (1.6.5). These two identities imply:

$$g_{\mu\sigma,\nu} + g_{\sigma\nu,\mu} - g_{\mu\nu,\sigma} = g_{\sigma\alpha}(\Gamma^{\alpha}_{\mu\nu} + \Gamma^{\alpha}_{\nu\mu}) + g_{\nu\alpha}(\Gamma^{\alpha}_{\sigma\mu} - \Gamma^{\alpha}_{\mu\sigma}) + g_{\mu\alpha}(\Gamma^{\alpha}_{\sigma\nu} - \Gamma^{\alpha}_{\nu\sigma}) = 2g_{\sigma\alpha}\Gamma^{\alpha}_{\mu\nu} \qquad \textbf{(1.6.6)}$$

where $g_{\sigma\mu,\nu} \equiv \dfrac{\partial g_{\sigma\mu}}{\partial y^{\nu}}$. By multiplying equation (1.6.6) with $g^{\beta\sigma}$ and summing over the index β, the following is revealed:

$$g^{\beta\sigma}(g_{\mu\sigma,\nu} + g_{\sigma\nu,\mu} - g_{\mu\nu,\sigma}) = 2g^{\beta\sigma}g_{\sigma\alpha}\Gamma^{\alpha}_{\mu\nu}$$

$$g^{\beta\alpha}(g_{\mu\sigma,\nu} + g_{\sigma\nu,\mu} - g_{\mu\nu,\sigma}) = 2\delta^{\beta}_{\alpha}\Gamma^{\alpha}_{\mu\nu}$$

$$\Gamma^{\beta}_{\mu\nu} = \frac{1}{2} g^{\beta\alpha}(g_{\mu\sigma,\nu} + g_{\sigma\nu,\mu} - g_{\mu\nu,\sigma}) \qquad \textbf{(1.6.7)}$$

The relation (1.6.7) derived here is a very important relation, since it states that we do not need the basis vectors to calculate the covariant derivatives, in fact we only need the metric components to do so. This relation will be very useful to us since the metric will later prove to contain all the significant physical information, especially the information about how gravitational waves manifest themselves.

1.7 Two-dimensional surfaces imbedded in E ^{3}

Anyone that has the slightest knowledge of the theory of general relativity knows that gravity can be pictured with the help of some abstract curved four-dimensional hyper surface. In order to move on towards our goal, which is to present the gravitational wave theory, we now must make this picture a lot clearer.

One of the most intuitive ways of looking at a curved surface is to place the surface in a space of higher dimension than that of the surface, where the extra dimensions of the ambient space provide the surface with the freedom to curve. A surface that is described in such a way is in mathematical terms called an imbedded surface.

Even though the human mind is only capable of visualizing two-dimensional surfaces imbedded in the Euclidian space of E^3, there is nothing preventing us from defining a hyper surface of dimension > 2 as an imbedded surface. In this section, we will try to describe the problems that appear when one tries to define some physical aspect within an imbedded surface. For the sake of simplicity we will here use a two-dimensional surface imbedded in E^3 as our illustrating example.

A two-dimensional curved surface S can be described as a continuous mapping from E ^{2} into E ^{3}, i.e. $\bar{f} : U \subset E^2 \to E^3$.

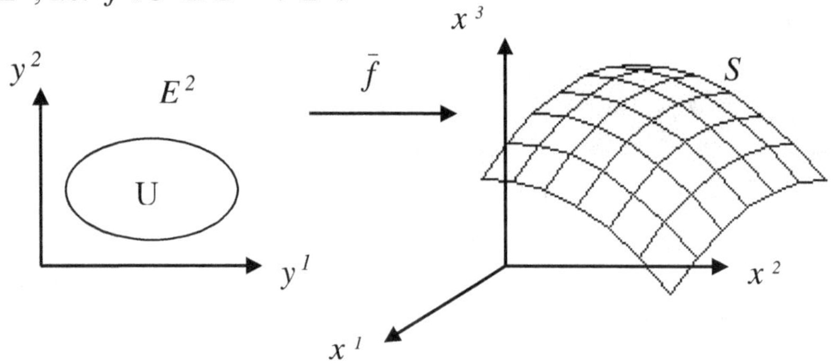

(figure 1)

At each point $p \in S$ there exists a tangentspace that is spanned by the two vectors $\{\vec{e}_1(p), \vec{e}_2(p)\}$. These two vectors constitute the basis for the tangentspace at p, defined by

$$\vec{e}_1(p) = \left.\frac{\partial \bar{f}}{\partial y^1}\right|_p \quad ; \quad \vec{e}_2(p) = \left.\frac{\partial \bar{f}}{\partial y^2}\right|_p \qquad \textbf{(1.7.1)}$$

The metric of the surface, is then given by these basis vectors by the expression

$$g_{\alpha\beta} = \vec{e}_\alpha \cdot \vec{e}_\beta \qquad \textbf{(1.7.2)}$$

The derivatives of these basis vectors with respect to the coordinates y^α will then be

$$\frac{\partial \vec{e}_\mu}{\partial y^\alpha} = \Gamma^\sigma_{\mu\alpha} \vec{e}_\sigma + b_{\mu\alpha} \vec{N} \qquad \textbf{(1.7.3)}$$

where the indexes α, σ, μ run from 1 to 2, and where $\vec{N}$ is the normal to the surface and $b_{\mu\alpha}$ is a continuous function not identical to zero[1].

This means that if we try to calculate the derivative of a vector $\vec{v} = v^{\alpha}\vec{e}_{\alpha}$ defined within some tangentplane of the surface S , the result of our derivation will at some point on the surface not lie within the tangentplane, i.e. the vector $\dfrac{\partial \vec{v}}{\partial y^{\alpha}}$ will "point out" from the surface.

This implies that if we try to model some physical aspect within some 2-dimensional curved surface, the derivatives of the vectors defined in this model will "jump out" of the physical reality we try to describe into the metaphysical ambient space, in which the surface is imbedded. In our case the aspect we will try to model is gravity, and the surface on which we want to model this aspect is the four-dimensional "hyper surface" of spacetime. It is then clear ,even though our example is based on only two-dimensional surfaces, that we can not describe the theory of gravity with the help of 4-dimensensional "hypersurfaces"imbedded in some ambient space, since this would demand the use of non physical extrinsic properties.

1.8 Intrinsic geometry

With de discussion of section 1.7 in mind, it is only natural to ask ourselves how we can describe the properties of a curved surface, or rather curved spaces, without imbedding the surface in a space of higher dimension? An answer to this question, concerning two-dimensional surfaces, was given by Gauss who showed that the intrinsic properties of a two-dimensional surface depend only on the metric defined on the surface, and the derivatives of the metric. Gauss did this by proving his Theorema Egregium[2], which states that the curvature K of a two-dimensional surface is given by:

$$K = \frac{g_{12,12} - \frac{1}{2}(g_{11,22} + g_{22,11})}{g_{11}g_{22} - (g_{12})^2} \qquad \textbf{(1.8.1)}$$

So even though we previously defined the curved surface and its metric through the imbedding $\bar{f} : U \subset E^2 \to E^3$, the properties of the surface can now, thanks to Theorema Egregium, be uniquely defined without the imbedding, as long as the metric of the surface is known. It is also important to stress the fact that this also means that the metric of the surface does not have to be defined through the imbedding, as it was previously done by (1.7.2).

The next step was taken by Riemann, who inspired by Gauss generalized the theory so it would be valid for "surfaces"(manifolds) of arbitrary dimension.

1.9 Manifolds and the local flattness theorem

Riemann invented the concept of a differentiable manifold. In our case we will concentrate on 4-dimesional differentiable manifolds, since it is these manifolds that are the main ingredients in the theory of general relativity. In our case we will use the following definition:

Definition of a 4-dimesional differentiable manifold

A 4-dimensional manifold is the set of points in R^4, together with a coordinate system, i.e. coordinates x^{α} with coresponding basis vektors $\vec{e}^{\alpha}$, and a symetric tensor $\overline{\overline{g}}$ defined on R^4, which components $g_{\alpha\beta}$ are continuous and differentiable functions of the coordinates x^{α}, from R^4 into R.

It is also important to make clear the concept of a curved manifold. A curved manifold is a manifold in which one never can find coordinates such that the metric becomes identical to the flat metric $\eta_{\alpha\beta}$ on the entire manifold, where the flat metric in general relativity is defined by:

$$\eta_{\alpha\beta} = \begin{pmatrix} 1 & 0 & 0 & 0 \\ 0 & -1 & 0 & 0 \\ 0 & 0 & -1 & 0 \\ 0 & 0 & 0 & -1 \end{pmatrix} \quad \textbf{(1.9.1)}$$

Now a new problem will appear when we try to calculate the derivatives of vectors $\vec{v} = v^{\alpha}\vec{e}_{\alpha}$, defined within a curved manifold. This problem comes from the fact that the derivative of such a vector with respect to the coordinate x^{β}, is given by:

$$\frac{\partial \vec{v}}{\partial x^{\beta}} = v^{\alpha}{}_{,\beta}\vec{e}_{\alpha} + v^{\alpha}\frac{\partial \vec{e}_{\alpha}}{\partial x^{\beta}} \quad \textbf{(1.9.2)}$$

Where the source of the problem can be found in the factor $\frac{\partial \vec{e}_{\alpha}}{\partial x^{\beta}}$ of the second term in the expression above. Since we no longer are in the flat space of E^3, or E^4 for that matter, we can not without further consideration say that the derivatives of the basis vectors are linear combinations of each other , i.e. $\frac{\partial \vec{e}_{\alpha}}{\partial x^{\beta}} = \Gamma^{\sigma}_{\alpha\beta}\vec{e}_{\sigma}$, where the linear coefficients $\Gamma^{\sigma}_{\alpha\beta}$ (the Christoffel symbols) are given by (1.6.7), because this requires that we still are in the flat space of E^4. But the solution to this problem is given by the local flatness theorem.

The local flatness theorem.

For every point $\bar{p} \in \mathrm{R}^4$ belonging to a 4-dimensional differentiable manifold, and for every $\varepsilon > 0$, it is always possible to find coordinates x^{α} such that:

$$|\bar{p} - \bar{x}| < \varepsilon \;\Rightarrow\; g_{\alpha\beta}(\bar{x}) = \eta_{\alpha\beta} + O(|\bar{x} - \bar{p}|^2) \quad \textbf{(1.9.3)}$$

Or alternatively:

For every point $\bar{p} \in \boldsymbol{R^4}$ belonging to a 4-dimensional differentiable manifold, it is always possible to find coordinates x^{α} such that:

$$1.)\;\; g_{\alpha\beta}(\bar{p}) = \eta_{\alpha\beta} \qquad 2.)\;\; \left.\frac{\partial g_{\alpha\beta}}{\partial x^{\mu}}\right|_{\bar{x}=\bar{p}} = 0 \qquad 3.)\;\; \left.\frac{\partial^2 g_{\alpha\beta}}{\partial x^{\mu}\partial x^{\nu}}\right|_{\bar{x}=\bar{p}} \neq 0 \quad \textbf{(1.9.4)}$$

A proof of this theorem can be found in B.F Schutz [3].

The local flatnes theorem implies that the metric on a curved manifold, at every point, can locally be regarded as being the metric of a flat space. In other words, this means that even though globally the metric is not that of flat space, one can always, locally at every point $\bar{p}$ of the manifold, regard the metric as being the dotproduct of the basis vectors in flat space, i.e.

$$g_{\alpha\beta}(\bar{x}) = \vec{e}_\alpha \cdot \vec{e}_\beta + O(|\bar{x} - \bar{p}|^2) \qquad \textbf{(1.9.5)}$$

The final conclusion is then that we can, locally on the manifold, use the same methods that were used in the case of a flat space (sections 1.4-1.6) to derive the covariant derivatives and the Christoffel symbols. This means that on a differentiable manifold the covariant derivatives and the Christoffel symbols are given by[4]:

$$\nabla_\beta v^\alpha = v^\alpha{}_{;\beta} \equiv v^\alpha{}_{,\beta} + v^\mu \Gamma^\alpha_{\mu\beta} \qquad \textbf{(1.9.6)}$$

$$\nabla_\beta v_\alpha = v_{\alpha;\beta} \equiv v_{\alpha,\beta} - v_\mu \Gamma^\mu_{\alpha\beta} \qquad \textbf{(1.9.7)}$$

$$T_{\alpha\beta;\mu} = T_{\alpha\beta,\mu} - \Gamma^\sigma_{\alpha\mu} T_{\sigma\beta} - \Gamma^\sigma_{\beta\mu} T_{\alpha\sigma} \qquad \textbf{(1.9.8)}$$

$$T^\beta_{\alpha;\mu} = T^\beta_{\alpha,\nu} + \Gamma^\beta_{\sigma\mu} T^\sigma_\alpha - \Gamma^\sigma_{\alpha\mu} T^\beta_\sigma \qquad \textbf{(1.9.9)}$$

$$T^\beta_{\alpha;\mu} = T^\beta_{\alpha,\nu} + \Gamma^\beta_{\sigma\mu} T^\sigma_\alpha - \Gamma^\sigma_{\alpha\mu} T^\beta_\sigma \qquad \textbf{(1.9.10)}$$

$$\Gamma^\beta_{\mu\nu} = \frac{1}{2} g^{\beta\alpha} (g_{\mu\sigma,\nu} + g_{\sigma\nu,\mu} - g_{\mu\nu,\sigma}) \qquad \textbf{(1.9.11)}$$

1.10 Parallel -transport and geodesics

The parallel transport of a vector is one of the most important ideas used in flat space-time physics. So when one tries to build a new physical theory defined within curved space-time, it is only natural to try and adopt this idea into the new physical theory. In this section, we will redefine the concept of parallel transport in order to make it useful in a curved space-time reality. We will also introduce the definition of a geodesic.

Definition of parallel-transport

A vector $\vec{v} = v^\alpha \vec{e}_\alpha$ is said to be parallel-transported along a curve $x^\alpha = x^\alpha(\tau)$, if

$$\nabla_{\bar{U}} \vec{v} \equiv v^\alpha_{;\beta} U^\beta \vec{e}_\alpha = 0 \qquad \textbf{(1.10.1)}$$

where $U^\beta = \dfrac{dx^\beta}{d\tau}$. Observe that the difference between parallel transport in a flat space and in a curved space is that in a flat space the condition for parallel transport is that the vectors must remain unchanged relative to a global coordinate system, whereas in a curved space, the condition is that the vectors must remain unchanged relative to a local coordinate system. It is in fact impossible to find a global coordinate system in a curved space.

Definition of a geodesic

A curve $x^{\alpha} = x^{\alpha}(\tau)$ is said to be a geodesic if it parallel-transports its own tangent, i.e.

$$\nabla_{\overline{U}} \overline{U} = 0 \qquad \textbf{(1.10.2)}$$

or equivalently:

$$\frac{d^2 x^{\alpha}}{d\tau^2} + \Gamma^{\alpha}_{\mu\nu} \frac{dx^{\mu}}{d\tau} \frac{dx^{\nu}}{d\tau} = 0 \qquad \textbf{(1.10.2)}$$

The concept of a geodesic will later prove to be of crucial importance to us, since in general relativity free particles are postulated to move along geodesics, defined on the four-dimensional manifold of space-time.

It can also be shown that the shortest path between two points on a manifold, is defined by a geodesic. So one way of looking at geodesics in general relativity, is to view them as the straight lines connecting different events in space-time

1.11 The Riemann curvature tensor

Since gravity in general relativity is described with the help of four-dimensional *curved* manifolds, and since gravitational waves are described as periodic changes of the space-time *curvature*, it is now necessary for us to take a closer look at the concept of *curvature*, in order to be able to formulate a theory for gravitational waves.

When we now are working with four dimensions we can no longer adopt the definition of curvature, which Gauss developed for two dimensional manifolds, given by (1.8.2). We must thus develop a more general definition of curvature, which can be useful on manifolds of arbitrary dimension. The idea of this new definition of curvature is based on the fact that, vectors that are being parallel-transported on a curved manifold and along closed loops will not always be unchanged, even though the transport starts and ends at the same point on the manifold. Two examples of such a parallel-transport can be seen in the figures below. The left figure shows parallel-transport on a flat manifold, and the right one shows parallel-transport on a curved manifold (The common start- and end-point is denoted *A*).

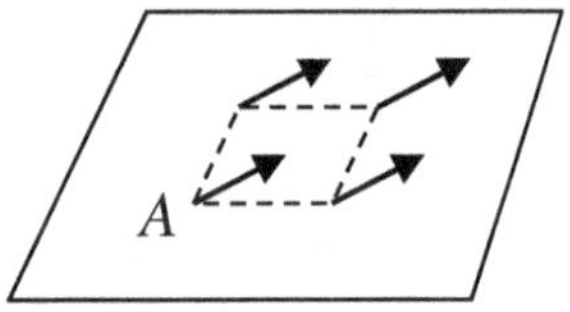

(figure 2a)

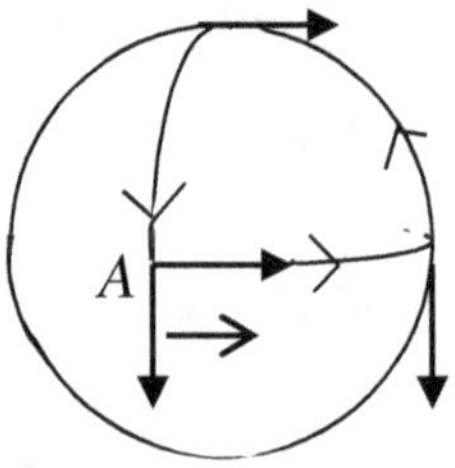

(figure 2b)

To do so, we will parallel-transport a vector around an infinitesimally small closed loop. The parallel-transport will start and end at the same point, and after it is done, the change in the components of the vector will be calculated.

The loop mentioned will here be constructed of the four curves C_1, C_2, C_3 and C_4 given by:

$$C_{AB} = \left\{ \bar{x} \;\middle|\; \bar{x} \in R^n,\ x^\sigma = b,\ a \le x^\rho \le a + \delta a \right\}$$
$$C_{BC} = \left\{ \bar{x} \;\middle|\; \bar{x} \in R^n,\ b \le x^\sigma \le b + \delta b,\ x^\rho = a + \delta a \right\}$$
$$C_{CD} = \left\{ \bar{x} \;\middle|\; \bar{x} \in R^n,\ x^\sigma = b + \delta b,\ a \le x^\rho \le a + \delta a \right\}$$
$$C_{DA} = \left\{ \bar{x} \;\middle|\; \bar{x} \in R^n,\ b \le x^\sigma \le b + \delta b,\ x^\rho = a \right\}$$

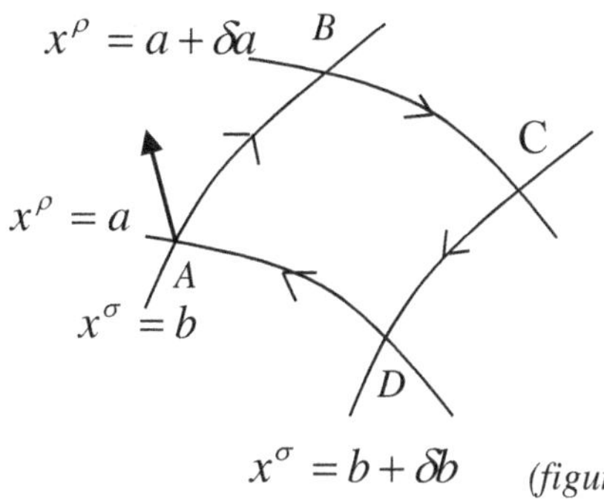

(figure3)

Where σ and ρ are fixed indices, such that $\sigma \neq \rho$.

Parallel-transport of the vector $\vec{v} = v^\alpha \vec{e}_\alpha$ from A to B along the curve C_{AB} gives:

$$\frac{dv^\alpha}{dx^\rho} = \frac{\partial v^\alpha}{\partial x^\rho} + v^\mu \Gamma^\alpha_{\mu\rho} = 0 \quad \Rightarrow \quad \frac{\partial v^\alpha}{\partial x^\rho} = -v^\mu \Gamma^\alpha_{\mu\rho} \qquad \textbf{(1.11.1)}$$

$$\Rightarrow$$

$$v^\alpha(B) - v^\alpha(A^{Start}) = -\int_a^{a+\delta a} v^\mu \Gamma^\alpha_{\mu\rho}\Big|_{x^\sigma = b} dx^\rho \qquad \textbf{(1.11.2)}$$

In the same manner one obtains through the parallel-transport of the vector along the remaining curves C_{AC}, C_{OD} and C_{DA}, the expressions:

$$v^\alpha(C) - v^\alpha(B) = -\int_b^{b+\delta b} v^\mu \Gamma^\alpha_{\mu\sigma}\Big|_{x^\rho = a+\delta a} dx^\sigma \qquad \textbf{(1.11.3)}$$

$$v^\alpha(D) - v^\alpha(C) = \int_a^{a+\delta a} v^\mu \Gamma^\alpha_{\mu\rho}\Big|_{x^\sigma = b+\delta b} dx^\rho \qquad \textbf{(1.11.4)}$$

$$v^\alpha(A^{End}) - v^\alpha(D) = \int_b^{b+\delta b} v^\mu \Gamma^\alpha_{\mu\sigma}\Big|_{x^\rho = a} dx^\sigma \qquad \textbf{(1.11.5)}$$

If we now define the difference between the vectors at the start and the end of the parallel-transport as $\delta v^a = v^\alpha(A^{End}) - v^\alpha(A^{start})$, it is easy to se that (1.11.1)-(1.11.4) imply:

$$\delta v^\alpha = \int_a^{a+\delta a} (v^\mu \Gamma^\alpha_{\mu\rho}\Big|_{x^\sigma = b+\delta b} - v^\mu \Gamma^\alpha_{\mu\rho}\Big|_{x^\sigma = b}) dx^\rho + \int_b^{b+\delta b} (v^\mu \Gamma^\alpha_{\mu\sigma}\Big|_{x^\rho = a} - v^\mu \Gamma^\alpha_{\mu\sigma}\Big|_{x^\rho = a+\delta a}) dx^\sigma \qquad \textbf{(1.11.6)}$$

Since δa and δb are very small, we can with the help of the *mean value theorem* do the following approximation on (1.11.6)

$$\delta v^{\alpha} \approx \int_{a}^{a+\delta a} \delta b \frac{\partial}{\partial x^{\sigma}}\left(v^{\mu}\Gamma^{\alpha}_{\mu\rho}\right)dx^{\rho} - \int_{b}^{b+\delta b} \delta a \frac{\partial}{\partial x^{\rho}}\left(v^{\mu}\Gamma^{\alpha}_{\mu\sigma}\right)dx^{\sigma} \approx \delta a \delta b\left[\frac{\partial}{\partial x^{\sigma}}\left(v^{\mu}\Gamma^{\alpha}_{\mu\rho}\right) - \frac{\partial}{\partial x^{\rho}}\left(v^{\mu}\Gamma^{\alpha}_{\mu\sigma}\right)\right]$$

(1.11.7)

Finally (1.11.7) together with (1.11.1) imply that the parallel-transport of the vector $\vec{v} = v^{\alpha}\vec{e}_{\alpha}$ will induce the following change in the components of the vector

$$\delta v^{\alpha} \approx \delta a \delta b(\Gamma^{\alpha}_{\lambda\rho,\sigma} - \Gamma^{\alpha}_{\lambda\sigma,\rho} + \Gamma^{\mu}_{\lambda\rho}\Gamma^{\alpha}_{\mu\sigma} - \Gamma^{\mu}_{\lambda\sigma}\Gamma^{\alpha}_{\mu\rho})v^{\lambda} \qquad \textbf{(1.11.8)}$$

It is now not very hard to see that the expression in the parenthesis of (1.11.8) defines the components of a tensor, which is covariant of order 3, and contravariant of order 1.

This tensor is called the *Riemann curvature tensor*, and it is defined by

$$R^{\alpha}{}_{\lambda\sigma\rho} = \Gamma^{\alpha}_{\lambda\rho,\sigma} - \Gamma^{\alpha}_{\lambda\sigma,\rho} + \Gamma^{\mu}_{\lambda\rho}\Gamma^{\alpha}_{\mu\sigma} - \Gamma^{\mu}_{\lambda\sigma}\Gamma^{\alpha}_{\mu\rho} \qquad \textbf{(1.11.9)}$$

So, with the help of the Riemann curvature tensor we have found a way of describing curvature of manifolds of arbitrary dimensions. To clarify this idea, we go back to our original example of the parallel transport of a vector v^{α} along an infinitesimally small loop located on a manifold. By studying this example we saw that the vector components v^{α} changed by

$$\delta v^{\alpha} = R^{\alpha}{}_{\lambda\sigma\rho} v^{\lambda} \delta a \delta b$$

This change is a measure of the curvature of the manifold, provided by the Riemann tensor. This measure gives us information about the curvature of the manifold, at the point on the manifold that is surrounded by the loop.

Some properties of the Riemann tensor

The local flatness theorem states, that at every point on a manifold it is always possible to find a local coordinate system, such that $g_{\alpha\beta,\mu} = 0$ (such a coordinate system is often referred to as a local inertial system). If we in this coordinate system use (1.9.11) and (1.11.9) to express the curvature tensor, we will get the following result:

$$R^{\lambda}{}_{\beta\mu\nu} = \Gamma^{\lambda}_{\beta\nu,\mu} - \Gamma^{\lambda}_{\beta\mu,\nu} = \frac{g^{\lambda\sigma}}{2}\left[g_{\sigma\nu,\beta\mu} - g_{\beta\nu,\sigma\mu} + g_{\beta\mu,\sigma\nu} - g_{\sigma\mu,\beta\nu}\right] \qquad \textbf{(1.11.10)}$$

By multiplying (1.11.10) with $g_{\alpha\lambda}$ and summing over the index λ we get the expression:

$$R_{\alpha\beta\mu\nu} = \frac{1}{2}\left[g_{\alpha\nu,\beta\mu} - g_{\beta\nu,\alpha\mu} + g_{\beta\mu,\alpha\nu} - g_{\alpha\mu,\beta\nu}\right] \qquad \textbf{(1.11.11)}$$

Then (1.11.11) together with the fact that partial derivatives commute implies that:

$$R_{\alpha\beta\mu\nu} = -R_{\beta\alpha\mu\nu} = -R_{\alpha\beta\nu\mu} = R_{\mu\nu\alpha\beta} \qquad \textbf{(1.11.12)}$$

$$R_{\alpha\beta\mu\nu} + R_{\alpha\nu\beta\mu} + R_{\alpha\mu\nu\beta} = 0 \qquad \textbf{(1.11.13)}$$

Since (1.11.12) and (1.11.12) are tensorial relations, they do not only apply in the local inertial system, they will apply in any coordinate system.

1.12 The geodesic deviation

Later in this chapter we will see that in general relativity, bodies move along geodesics in the curved manifold of space-time. These geodesics are the straight lines of space-time, in which no acceleration relative to the co-moving coordinate system can be measured. This means that in general relativity the concept of acceleration must be redefined. This new definition of acceleration will be based on the fact that two parallel distinct geodesics, on a curved manifold, will not always remain parallel when they are extended.

Lets consider two geodesics, $x^{\alpha} = x^{\alpha}(\tau)$ and $x^{\alpha'} = x^{\alpha'}(\tau')$ who are parallel when $\tau = \tau_A$ and $\tau' = \tau'_A$. Let the initial separation of the geodesics be defined by $\xi^{\alpha} = x^{\alpha'}(\tau'_A) - x^{\alpha}(\tau_A)$, and let this separation be small, i.e. $|\xi^{\alpha}| << 1$.

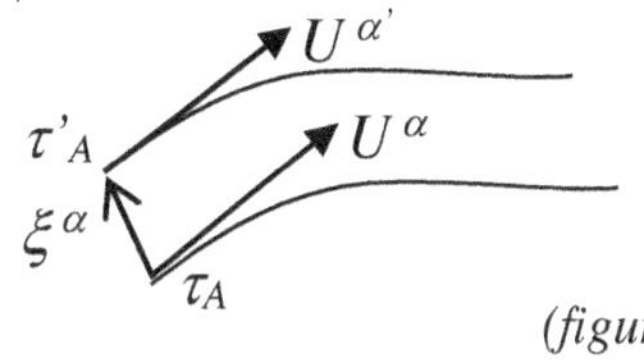

(figure 4)

The geodesic equation (1.10.2) then implies that:

$$\left.\frac{d^2x^{\alpha}}{d\tau^2}\right|_{\tau=\tau_A} = -\left.\frac{dx^{\sigma}}{d\tau}\frac{dx^{\beta}}{d\tau}\Gamma^{\alpha}_{\sigma\beta}\right|_{\tau=\tau_A} \quad ; \quad \left.\frac{d^2x^{\alpha'}}{d\tau'^2}\right|_{\tau'=\tau'_A} = -\left.\frac{dx^{\sigma'}}{d\tau'}\frac{dx^{\beta'}}{d\tau'}\Gamma^{\alpha'}_{\sigma'\beta'}\right|_{\tau'=\tau'_A} \qquad \textbf{(1.12.1)}$$

Furthermore (1.12.1) implies:

$$\frac{d^2\xi^{\alpha}}{d\tau^2} = \left.\frac{dx^{\sigma'}}{d\tau'}\frac{dx^{\beta'}}{d\tau'}\Gamma^{\alpha'}_{\sigma'\beta'}\right|_{\tau'=\tau'_A} - \left.\frac{dx^{\sigma}}{d\tau}\frac{dx^{\beta}}{d\tau}\Gamma^{\alpha}_{\sigma\beta}\right|_{\tau=\tau_A} \qquad \textbf{(1.12.2)}$$

Now the *mean value theorem* together with the assumption that $|\xi^{\alpha}| << 1$ makes it possible to do the following approximation:

$$\frac{d^2\xi^{\alpha}}{d\tau^2} \approx -\left(\frac{dx^{\sigma}}{d\tau}\frac{dx^{\beta}}{d\tau}\Gamma^{\alpha}_{\sigma\beta}\right)_{,\mu}\xi^{\mu} = -\frac{dx^{\sigma}}{d\tau}\frac{dx^{\beta}}{d\tau}\Gamma^{\alpha}_{\sigma\beta,\mu}\xi^{\mu} = -U^{\sigma}U^{\beta}\Gamma^{\alpha}_{\sigma\beta,\mu}\xi^{\mu} \qquad \textbf{(1.12.3)}$$

where $U^{\sigma} = \dfrac{dx^{\sigma}}{d\tau}$.

In our forthcoming derivations we will need the initial parallelism condition [5] asserting that the geodesics begin parallel: $\xi^{\alpha}{}_{,\mu} = 0$ **(1.12.4)**

This condition immediately implies the useful relation:

$$\frac{d^2\xi^{\alpha}}{d\tau^2} = \left(\xi^{\alpha}{}_{,\beta}U^{\beta}\right)_{,\mu}U^{\mu} = \xi^{\alpha}{}_{,\beta\mu}U^{\beta}U^{\mu} \qquad \textbf{(1.12.5)}$$

We now have all the tools needed for calculating the equation of geodesic deviation, which is defined by:

$$\frac{D^2\xi^{\alpha}}{d\tau^2} \equiv \nabla_{\bar{U}}\nabla_{\bar{U}}\xi^{\alpha} \qquad \textbf{(1.12.6)}$$

The last step is now to express the right-hand side of (1.12.6) more explicitly.

$$\nabla_{\bar{U}}\nabla_{\bar{U}}\xi^{\alpha} = \nabla_{\bar{U}}\left[\left(\xi^{\alpha}{}_{,\beta} + \xi^{\sigma}\Gamma^{\alpha}_{\sigma\beta}\right)U^{\beta}\right] = \nabla_{\bar{U}}\left(\xi^{\alpha}{}_{,\beta} + \xi^{\sigma}\Gamma^{\alpha}_{\sigma\beta}\right)U^{\beta} \qquad \textbf{(1.12.7)}$$

Since $\left(\xi^{\alpha}{}_{,\beta} + \xi^{\sigma}\Gamma^{\alpha}_{\sigma\beta}\right)$ is a tensor, which is covariant of order 1 and contravariant of order 1, we can use (1.9.9) to calculate the covariant derivative for this expression, i.e.

$$\nabla_{\bar{U}}\nabla_{\bar{U}}\xi^{\alpha} = \frac{d^2\xi^{\alpha}}{d\tau^2} + \xi^{\sigma}\Gamma^{\alpha}_{\sigma\beta,\mu}U^{\mu}U^{\beta} + \xi^{\sigma}\Gamma^{\nu}_{\sigma\beta}\Gamma^{\alpha}_{\nu\mu}U^{\mu}U^{\beta} - \xi^{\sigma}\Gamma^{\alpha}_{\sigma\nu}\Gamma^{\nu}_{\beta\mu}U^{\mu}U^{\beta} \qquad \textbf{(1.12.8)}$$

Observe that we also used (1.12.4) and (1.12.5) in the derivation of (1.12.8).
Now (1.12.3) helps us to rewrite (1.12.8) in the following way:

$$\nabla_{\bar{U}}\nabla_{\bar{U}}\xi^{\alpha} = \left(\Gamma^{\alpha}_{\beta\sigma,\mu} - \Gamma^{\alpha}_{\beta\mu,\sigma} + \Gamma^{\nu}_{\beta\sigma}\Gamma^{\alpha}_{\nu\mu} - \Gamma^{\nu}_{\mu\beta}\Gamma^{\alpha}_{\nu\sigma}\right)U^{\beta}U^{\mu}\xi^{\sigma} \qquad \textbf{(1.12.9)}$$

But the expression in the parenthesis of (1.12.9) is nothing but the components of the Riemann curvature tensor, and thus the equation of geodesic deviation becomes:

$$\frac{D^2\xi^{\alpha}}{d\tau^2} \equiv \nabla_{\bar{U}}\nabla_{\bar{U}}\xi^{\alpha} = R^{\alpha}{}_{\beta\mu\sigma}U^{\beta}U^{\mu}\xi^{\sigma} \qquad \textbf{(1.12.10)}$$

The equation of geodesic deviation will be of great importance to us, since it will help us to complete the transition from Newtonian gravitational theory to general relativity, and because it will prove to be the equation that describes how gravitational waves interact with material particles.

1.13 The Einstein tensor

In order to formulate the equations that describe how the four-dimensional manifold of space-time is shaped by the presence of mass, or rather energy, we need to develop a new geometrical quantity called the Einstein tensor. This will be done by examining the properties of the Riemann tensor.

The Bianchi identities

The expression (1.11.11) tells us that in a locally inertial coordinate system, the covariant derivative of $R_{\alpha\beta\mu\nu}$ can be expressed as:

$$R_{\alpha\beta\mu\nu;\lambda} = \frac{1}{2}\left(g_{\alpha\nu,\beta\mu\lambda} - g_{\alpha\mu,\beta\nu\lambda} + g_{\beta\mu,\alpha\nu\lambda} - g_{\beta\nu,\alpha\mu\lambda}\right) \quad \textbf{(1.13.1)}$$

From this expression, the symmetry of the metric tensor and the fact that partial derivatives commute, it is easy to show that

$$R_{\alpha\beta\mu\nu;\lambda} + R_{\alpha\beta\lambda\mu;\nu} + R_{\alpha\beta\nu\lambda;\mu} = 0 \quad \textbf{(1.13.2)}$$

Since (1.13.2) is a tensorial relation, it is not only valid in a local inertial coordinate system, it is also valid in any coordinate system. These identities are called the *Bianchi identities*.

The Ricci tensor

The Ricci tensor is a symmetric tensor defined by $R_{\alpha\beta} \equiv R^{\mu}{}_{\alpha\mu\beta}$ **(1.13.3)**

The Ricci scalar

The Ricci scalar is defined by $R \equiv g^{\alpha\beta} R_{\alpha\beta} = g^{\alpha\beta} R^{\mu}{}_{\alpha\mu\beta}$ **(1.13.4)**

The Einstein tensor

To derive the Einstein tensor we begin by multiplying the Bianchi identities (1.13.2) with $g^{\mu\alpha}$ and sum over the index α, the result of this operation is

$$R_{\beta\nu;\lambda} - R_{\beta\lambda;\nu} + R^{\mu}{}_{\beta\nu\lambda;\mu} = 0 \quad \textbf{(1.13.5)}$$

Observe that we have used the fact that $g_{\alpha\beta;\mu} = 0$ in this derivation, and will continue to do so in the following derivations. The next step is to multiply (1.13.5) with $g^{\beta\nu}$ and sum over the index β, the result is:

$$R_{;\lambda} - 2R^{\mu}{}_{\lambda;\mu} = 0$$

$$\Leftrightarrow$$

$$\left(R^{\mu}{}_{\lambda} - \frac{\delta^{\mu}_{\lambda}}{2} R\right)_{;\mu} = 0 \quad \textbf{(1.13.6)}$$

Finally we multiply (1.13.6) with $g^{\nu\lambda}$ and sum over the index λ, which gives the result

$$\left(R^{\mu\nu} - \frac{g^{\mu\nu}}{2} R\right)_{;\mu} = 0 \quad \textbf{(1.13.7)}$$

The expression in the parenthesis of (1.13.7) defines a new symmetric tensor, called the Einstein tensor, i.e. the Einstein tensor is defined by

$$G^{\mu\nu} = R^{\mu\nu} - \frac{g^{\mu\nu}}{2} R \qquad \textbf{(1.13.8)}$$

$$G^{\mu\nu}{}_{;\mu} = 0 \qquad \textbf{(1.13.9)}$$

The property (1.3.9) of the Einstein tensor is the main reason for developing this tensor. It is this property that later on will enable us to formulate the physically relevant equations of general relativity.

2 General relativity

In this chapter some aspects of the theory of general relativity will be discussed. This is done in order to make it possible for those readers, who have not or just briefly studied the subject, to understand how the theory of gravitational waves has been developed.

2.1 The postulates of general relativity

The main idea of the theory of general relativity is that gravity not is modelled by forces between massive particles as it is done in Newtonian theory, instead gravity is modelled as an geometrical aspect of space and time, where the presence of energy is the source of this aspect. The ideas that gave birth to the theory are as follows

I. Space-time, i.e. the set of all events, is a 4-dimensional manifold with a metric $g_{\mu\nu}$.

II. The metric is determined by "rulers" and chronometers. The *distance* dS between two nearby events in space-time is defined to be:

$$dS \equiv \sqrt{\left|g_{\mu\nu}dx^{\mu}dx^{\nu}\right|} \qquad \textbf{(2.1.1)}$$

The *proper time* $d\tau$ separating two events which close together in time and taking place in the same spatial point in space-time, is defined by:

$$d\tau \equiv \sqrt{g_{00}}\,dt \qquad \textbf{(2.1.2)}$$

III. The metric of space-time can always locally be expressed in the Minkowski–form

$$\eta_{\mu\nu} = \begin{pmatrix} 1 & 0 & 0 & 0 \\ 0 & -1 & & 0 \\ 0 & 0 & -1 & 0 \\ 0 & 0 & 0 & -1 \end{pmatrix}$$

at every specific event, by choosing the appropriate coordinates. The coordinates in which the metric takes the Minkowski–form, will be defined to be the *local inertial fram of general relativity*.

IV. *The weak equivalence principle*

Freely falling bodies move along time-like geodesics, where a time-like geodesic is a curve $x^{\alpha} = x^{\alpha}(\tau)$ in space-time, which apart from being a geodesic also have the property:

$$g_{\mu\nu}\frac{dx^{\mu}}{d\tau}\frac{dx^{\upsilon}}{d\tau} > 0 \qquad \textbf{(2.1.3)}$$

V. *The strong equivalence principle*

Every physical law that can be expressed in tensor notation in special relativity have the exact same form in the local inertial frame of general relativity. This means that physical laws in special relativity, of the form mentioned above, can be transformed to be valid in general relativity by replacing all partial derivatives with covariant derivatives.

Example:

In special relativity the conservation of four-momentum in a perfect fluid is expressed by

$$T^{\mu\nu}{}_{,\nu} = 0 \tag{2.1.4}$$

,where $T^{\mu\nu}$ is the energy-momentum tensor of the fluid, which we will discuss in greater detail in the next section.

The strong equivalence principle then implies that this law in general relativity takes the more general form:

$$T^{\mu\nu}{}_{;\nu} = 0 \tag{2.1.5}$$

2.2 The Einstein equations

In Newtonian theory, gravity can only exist where there exists matter. However Einstein showed that matter and energy are only different faces of the same coin. This encouraged him to make the conclusion that gravity is not only created by the presence of matter, it is in fact the product of the presence of energy.

The next question we want to answer is how does the presence of energy shape the manifold of space-time? Or equivalently, which form do the equations have, that carry the information about how the metric is related to the presence of energy?

The answer to this question is of course given by Einstein's equations. In this section we will try to "motivate" the shape of these equations[8]. We will also only consider the case when matter is the singular energy-form present to create gravity.

The first step in our motivation of Einstein's equations is to study the relative acceleration, in Newtonian theory, between two bodies that are freely falling towards a third more massive body. If the masses of the two falling bodies are negligible compared to the mass of the third body and the distance separating the bodies is given by ξ, then the relative acceleration is

$$\frac{d^2\bar{\xi}}{dt^2} = \nabla \cdot \Phi\big|_{\bar{r}_1} - \nabla \cdot \Phi\big|_{\bar{r}_2} \tag{2.2.1}$$

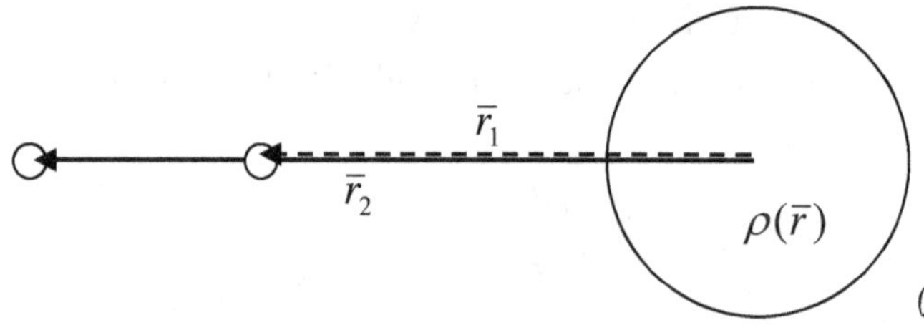

(figure 5)

where $\bar{\xi} = \bar{r}_2 - \bar{r}_1$, Φ is the Newtonian potential of the more massive body and where the indexes 1 and 2 correspond to the to less massive bodies.

If we assume that the distance between the less massive bodies is small, i.e. $|\bar{\xi}| << 1$, we can use the *mean value theorem* to make the following approximation:

$$\frac{d^2\bar{\xi}}{dt^2} = -(\nabla \cdot \Phi)_{,\beta}\xi^\beta \qquad \textbf{(2.2.2)}$$

If we also assume that the Newtonian potential Φ is spherically symmetric, we can rewrite (2.2.2) as follows:

$$\frac{d^2\xi^\alpha}{dt^2} = -(\nabla^2\Phi)\xi^\alpha \qquad \textbf{(2.2.3)}$$

But $\nabla^2\Phi = 4\pi G\rho_0$, where ρ_0 is the rest-mass-distribution of the more massive body and G is the gravitational constant. This means that the relative acceleration in Newtonian theory can be expressed as:

$$\frac{d^2\xi^\alpha}{dt^2} = -4\pi G\rho_0\xi^\alpha \qquad \textbf{(2.2.4)}$$

Now we continue to consider the same situation within the framework of general relativity. In this perspective the two less massive bodies move along geodesics in the manifold of space-time. This means that we can use the equation of geodesic deviation to describe the relative acceleration of the two bodies, i.e.

$$\frac{D^2\xi^\alpha}{d\tau^2} = R^\alpha{}_{\beta\mu\sigma}U^\beta U^\mu \xi^\sigma \qquad \textbf{(2.2.5)}$$

where U^β is the four-velocity of the bodies.

If we now compare (2.2.4) and (2.2.5) with each other, the following correspondence is revealed:

$$R^\alpha{}_{\beta\mu\alpha}U^\beta U^\mu \leftrightarrow -4\pi G\rho_0 \qquad \textbf{(2.2.6)}$$

Now the question that remains is how the source-term is expressed in general relativity? In Newtonian theory the source-term is simply expressed by the rest-mass-distribution ρ_0, but if we try to use the mass-energy distribution ρc^2 as the source-term in our new theory problems will arise. These problems are related to the fact that the mass-energy distribution is dependent on the choice of coordinate system. Thus, the choice of using the mass-energy-distribution as the source-term will make it impossible to define an invariant theory, in which the form of the equations describing gravity is independent of the choice of coordinate system[6]. Let us instead turn our attention to the energy-momentum tensor, which in special relativity is used to describe the energy- and momentum- properties of a perfect fluid. This tensor is defined by:

$$T^{\beta\mu} = \left(\rho + \frac{p}{c^2}\right)U^\beta U^\mu - p\eta^{\beta\mu} \qquad \textbf{(2.2.7)}$$

Where ρ and p are the mass density and the pressure of the fluid element in the rest system of the element, and where U^β is the four-velocity of the fluid-element.

Let us assume that the velocity v of the two less massive particles is small, i.e. $v << c$, then the following approximation can be done:

$$T_{\beta\mu}U^{\beta}U^{\mu} \approx c^4 \rho \qquad \textbf{(2.2.8)}$$

If we now take a good look at (2.2.6) and (2.2.8) we get the correspondence

$$R^{\alpha}{}_{\beta\mu\alpha}U^{\beta}U^{\alpha} \leftrightarrow -4\pi G\rho_0 \leftrightarrow -T_{\beta\mu}U^{\beta}U^{\mu} \qquad \textbf{(2.2.9)}$$

From this correspondence and the fact that $R^{\alpha}{}_{\beta\mu\alpha} = -R^{\alpha}{}_{\beta\alpha\mu} = -R_{\beta\mu}$, we take our first step towards our goal, and assume that the equations that we seek have the form

$$R_{\beta\mu} = \kappa T_{\beta\mu} \qquad \textbf{(2.2.10)}$$

where κ is a constant.

However it turns out that this assumption will not be altogether realistic. The naivety of equation (2.2.10) is the result of the conservation of energy and momentum relation

$$T^{\beta\mu}{}_{;\mu} = 0 \qquad \textbf{(2.2.11)}$$

Which together with (2.2.10) implies that

$$R^{\beta\mu}{}_{;\mu} = 0 \qquad \textbf{(2.2.12)}$$

Taking into account that the Einstein tensor satisfies the relation $G^{\beta\mu}{}_{;\mu} = 0$, we then get that

$$R_{;\mu} = 0 \qquad \textbf{(2.2.13)}$$

Then, it is not very hard to see that (2.2.10) and (2.2.13) give the result

$$T_{;\mu} = 0 \qquad \textbf{(2.2.13)}$$

i.e. the trace of the energy momentum tensor ($tr(T^{\beta\mu}) = T^{\beta}_{\beta}$) is constant through the whole universe, which is not a very physical assumption. But if we instead assume that the equations of gravity have the form

$$G^{\beta\mu} = \kappa T^{\beta\mu} \qquad \textbf{(2.2.14)}$$

the inconvenience of (2.2.13) is avoided, since (2.2.11) is fulfilled a priori by the Einstein tensor ($G^{\beta\mu}{}_{;\mu} = 0$). The equations (2.2.14) are called the *Einstein equations* and they are the fundament on which the theory of gravitational waves were built.

The constant κ

We now continue with determining the constant κ. This will be done by considering static weak gravitational fields (Newtonian gravitational fields) within the framework of general relativity and then comparing the conclusions of these considerations with the Newtonian theory

A weak gravitational field has a source of, which Newtonian potential and velocity are small, i.e.

$$|\Phi| << c^2 \quad \& \quad |\bar{v}| << c \tag{2.2.15}$$

It is easy to show in this weak field limit that [9]

$$|T^{00}| >> |T^{0k}| >> |T^{ik}| \tag{2.2.16}$$

$$T^{00} \approx \rho c^2 \tag{2.2.17}$$

We also rewrite the Einstein equations to a form that is more convenient to our forthcoming derivations:

$$R_{\mu\nu} = \kappa\left(T_{\mu\nu} - \frac{1}{2} g_{\mu\nu} T\right) \tag{2.2.18}$$

When working with weak gravitational fields the metric can be written on the form

$$g_{\mu\nu} = \eta_{\mu\nu} + h_{\mu\nu} \tag{2.2.19}$$

where $\eta_{\mu\nu}$ is the flat space Minkowski metric and $h_{\mu\nu}$ are small perturbations, satisfying the conditions:

$$|h_{\mu\nu}| << 1 \quad \& \quad |h_{\mu\nu,\alpha}| << 1 \tag{2.2.20}$$

The Christoffel symbols can now in a first order approximation (omitting all powers of $h_{\mu\nu}$ and $h_{\mu\nu,\alpha}$ greater than 1) be expressed as:

$$\Gamma^{\alpha}_{\mu\nu} = \frac{\eta^{\alpha\sigma}}{2}\left(h_{\mu\sigma,\nu} + h_{\sigma\nu,\mu} - h_{\mu\nu,\sigma}\right) \tag{2.2.21}$$

Similarly the Ricci tensor can be expressed in the first order approximation as

$$R_{\mu\nu} = \Gamma^{\alpha}_{\mu\nu,\alpha} - \Gamma^{\alpha}_{\mu\alpha,\nu} \tag{2.2.22}$$

This implies together with the assumption that the field is static that

$$R_{00} = \Gamma^{\alpha}_{00,\alpha} \tag{2.2.23}$$

But (2.2.21) and the static field assumption can be used to express R_{00} as

$$R_{00} = -\frac{1}{2} h_{00}{}^{,\alpha}{}_{,\alpha} = \frac{1}{2}\nabla^2 h_{00} \qquad \textbf{(2.2.24)}$$

We can now with the help of (2.2.16), (2.2.17) , (2.2.18) and (2.2.24) get the following approximate relation

$$\nabla^2 h_{00} \approx \kappa \rho c^2 \qquad \textbf{(2.2.25)}$$

Comparing this to the Newtonian relation $\nabla^2 \Phi = 4\pi G\rho$, we can make the identification

$$h_{00} = \frac{\kappa c^2}{4\pi G}\Phi \qquad \textbf{(2.2.26)}$$

The next step in our quest for the constant κ is to consider the geodesic equations for a particle that is falling freely towards the source

$$\frac{d^2 x^\mu}{d\tau^2} + \Gamma^\mu_{\alpha\beta}\frac{dx^\alpha}{d\tau}\frac{dx^\beta}{d\tau} = 0 \qquad \textbf{(2.2.27)}$$

Since the velocity of the particle relative to the source is small, i.e. $\left|\frac{dx^i}{d\tau}\right| \approx |\bar{v}| << c$, the following approximation can be made:

$$\frac{d^2 x^\mu}{d\tau^2} \approx -\Gamma^\mu_{00} c^2 \qquad \textbf{(2.2.27)}$$

With the help of (2.2.21) and the assumption that the field is static (2.2.27) can be rewritten as

$$\frac{d^2 x^\mu}{d\tau^2} \approx -\frac{c^2}{2}(\nabla \cdot h_{00})^\mu \qquad \textbf{(2.2.28)}$$

Comparing this to the Newtonian acceleration $\frac{d^2 x^\mu}{dt^2} = -\nabla \cdot \Phi$, we can make the identification

$$h_{00} \approx \frac{2}{c^2}\Phi \qquad \textbf{(2.2.29)}$$

If we compare (2.2.26) and (2.2.29) it then becomes clear that

$$\kappa = \frac{8\pi G}{c^4} \qquad \textbf{(2.2.30)}$$

3 Einstein's linearized equations for weak gravitational fields

In this chapter we will reformulate the Einstein equations so that they can be expressed as a wave equation. This is done by linearizing the equations by assuming that the gravitational field is weak, and by choosing appropriate coordinates in which to express the equations.

This chapter will be divided into two sections. In the first section the linearization of the Einstein equations will be discussed. In the second section it will be proved that one always can choose a set of coordinates, such that the linearized equations take on the form of an inhomogeneous wave equation.

3.1 The weak field approximation (linear approximation)

A weak gravitational field is a field corresponding to a metric in space-time that only slightly differs from the Minkowski metric (the flat metric). It is therefore realistic to assume that these fields have a metric of the form

$$g_{\alpha\beta} = \eta_{\alpha\beta} + h_{\alpha\beta} \qquad \textbf{(3.1.1)}$$

where

$$|h_{\alpha\beta}| << 1 \quad \& \quad |h_{\alpha\beta,\mu}| << 1 \qquad \textbf{(3.1.2)}$$

The Christoffel symbols in the linear approximation

The Christoffel symbols can with the aid of (1.9.11), (3.1.1) and (3.1.2) in the first order approximation (omitting all powers of $h_{\mu\nu}$ and $h_{\mu\nu,\alpha}$ greater than 1) be expressed as

$$\Gamma^{\mu}_{\alpha\beta} = \frac{1}{2}(h^{\mu}{}_{\alpha,\beta} + h^{\mu}{}_{\beta,\alpha} - h_{\alpha\beta}{}^{,\mu}) \qquad \textbf{(3.1.3)}$$

Where $h^{\mu}{}_{\beta,\alpha} \equiv \eta^{\mu\sigma} h_{\sigma\beta,\alpha}$. Observe that in the first order approximation we will use the Minkowski metric to lower and raise indexes.

The Riemann tensor in the linear approximation

We can now use (1.11.9) and (3.1.3) in order to express the Riemann tensor in the linear approximation

$$R^{\alpha}{}_{\beta\mu\nu} = \Gamma^{\alpha}_{\beta\nu,\mu} - \Gamma^{\alpha}_{\beta\mu,\nu} + \Gamma^{\sigma}_{\beta\nu}\Gamma^{\alpha}_{\sigma\mu} - \Gamma^{\sigma}_{\beta\mu}\Gamma^{\alpha}_{\sigma\nu} \approx \Gamma^{\alpha}_{\beta\nu,\mu} - \Gamma^{\alpha}_{\beta\mu,\nu} \quad \Rightarrow$$

$$R^{\alpha}{}_{\beta\mu\nu} = \frac{1}{2}\left(h^{\alpha}{}_{\nu,\beta\mu} + h_{\beta\mu}{}^{,\alpha}{}_{,\nu} - h^{\alpha}{}_{\mu,\beta\nu} - h_{\beta\nu}{}^{,\alpha}{}_{,\mu}\right) \qquad \textbf{(3.1.4)}$$

The Ricci tensor in the linear approximation

With the help of (3.1.4) the Ricci tensor can in the linear approximation be expressed as

$$R_{\beta\nu} = R^{\alpha}{}_{\beta\alpha\nu} = \frac{1}{2}\left(h^{\alpha}{}_{\nu,\beta\alpha} + h_{\beta\alpha}{}^{,\alpha}{}_{,\nu} - h^{\alpha}{}_{\alpha,\beta\nu} - h_{\beta\nu}{}^{,\alpha}{}_{,\alpha}\right) \quad \Rightarrow$$

$$R_{\beta\nu} = \frac{1}{2}(h^{\alpha}{}_{\nu,\beta\alpha} + h_{\beta\alpha}{}^{,\alpha}{}_{,\nu} - h_{,\beta\nu} - h_{\beta\nu}{}^{,\alpha}{}_{,\alpha}) \qquad \textbf{(3.1.5)}$$

where $h \equiv h^{\alpha}{}_{\alpha} = tr(h_{\alpha\beta})$

The Ricci scalar in the linear approximation

By using (3.1.5) the Ricci scalar can in the linear approximation be expressed as

$$R = g^{\beta\nu}R_{\beta\nu} \approx \eta^{\beta\nu}R_{\beta\nu} \approx \frac{\eta^{\beta\nu}}{2}(h^{\alpha}{}_{\nu,\beta\alpha} + h_{\beta\alpha}{}^{,\alpha}{}_{,\nu} - h_{,\beta\nu} - h_{\beta\nu}{}^{,\alpha}{}_{,\alpha}) \quad \Rightarrow$$

$$R = h_{\alpha\beta}{}^{,\alpha\beta} - h^{,\alpha}{}_{,\alpha} \qquad \textbf{(3.1.6)}$$

The Einstein tensor in the linear approximation

Consequently the Einstein tensor can with the help of (3.1.5) and (3.1.6) be given the form

$$G_{\mu\nu} = R_{\mu\nu} - \frac{g_{\mu\nu}}{2}R \approx R_{\mu\nu} - \frac{\eta_{\mu\nu}}{2}R \quad \Rightarrow$$

$$G_{\mu\nu} = \frac{1}{2}\left[h^{\alpha}{}_{\nu,\mu\alpha} + h_{\mu\alpha}{}^{,\alpha}{}_{,\nu} - h_{,\mu\nu} - h_{\mu\nu}{}^{,\alpha}{}_{,\alpha} - \eta_{\mu\nu}(h_{\alpha\beta}{}^{,\alpha\beta} - h^{,\alpha}{}_{,\alpha})\right] \qquad \textbf{(3.1.7)}$$

The trace reverse

In order to simplify the expression (3.1.7) for the Einstein tensor, we introduce the trace reverse of the metric perturbation, defined by:

$$\bar{h}^{\alpha\beta} = h^{\alpha\beta} - \frac{\eta^{\alpha\beta}}{2}h \qquad \textbf{(3.1.8)}$$

It is easy to show that this definition implies the two additional relations

$$\bar{h} = -h \qquad \textbf{(3.1.9)}$$

$$h^{\alpha\beta} = \bar{h}^{\alpha\beta} - \frac{\eta^{\alpha\beta}}{2}\bar{h} \qquad \textbf{(3.1.10)}$$

The Einstein tensor in the linear approximation expressed with the trace reverse

If we do the substitution defined by (3.1.9) and (3.1.10) in (3.1.7) we get

$$G_{\mu\nu} = \frac{1}{2}\Big[(\bar{h}^{\alpha}{}_{\nu} - \frac{\delta^{\alpha}{}_{\nu}}{2}\bar{h})_{,\mu\alpha} + (\bar{h}_{\mu\alpha} - \frac{\eta_{\mu\alpha}}{2}\bar{h})^{,\alpha}{}_{,\nu} + \bar{h}_{,\mu\nu} - (\bar{h}_{\mu\nu} - \frac{\eta_{\mu\nu}}{2}\bar{h})^{,\alpha}{}_{,\alpha} -$$
$$- \eta_{\mu\nu}(\bar{h}_{\alpha\beta} - \frac{\eta_{\alpha\beta}}{2}\bar{h})^{,\alpha\beta} - \eta_{\mu\nu}\bar{h}^{,\alpha}{}_{,\alpha}\Big] \quad \Rightarrow$$

$$G_{\mu\nu} = -\frac{1}{2}[\Box\bar{h}_{\mu\nu} + \eta_{\mu\nu}\bar{h}_{\alpha\beta}{}^{,\alpha\beta} - \bar{h}_{\mu\alpha}{}^{,\alpha}{}_{,\nu} - \bar{h}_{\alpha\nu,\mu}{}^{,\alpha}] \qquad \textbf{(3.1.11)}$$

here we have used the definition of the D'Alembertian operator

$$\bar{h}_{\mu\nu}{}^{,\alpha}{}_{,\alpha} = \eta^{\alpha\beta}\bar{h}_{\mu\nu,\beta\alpha} = \eta^{\alpha\beta}\frac{\partial^2\bar{h}_{\mu\nu}}{\partial x^{\beta}\partial x^{\alpha}} = \left(\frac{1}{c^2}\frac{\partial^2}{\partial t^2} - \nabla^2\right)\bar{h}_{\mu\nu} \equiv \Box\bar{h}_{\mu\nu} \qquad \textbf{(3.1.12)}$$

3.2 The gauge transformation and Einstein's linearized equations

Taking a good look at the expression (3.1.11) for the Einstein tensor, it is easy to see that this expression would be simplified considerably if

$$\bar{h}^{\mu\nu}{}_{,\nu} = 0 \qquad \textbf{(3.2.1)}$$

If these additional conditions are satisfied, the Einstein tensor would be expressed by

$$G_{\mu\nu} = -\frac{1}{2}\Box\bar{h}_{\mu\nu} \qquad \textbf{(3.2.2)}$$

So, is it possible for the condition (3.2.1) to be satisfied? To answer this question we must first consider the Einstein equations

$$G_{\mu\nu} = \kappa T_{\mu\nu} \qquad \textbf{(3.2.3)}$$

These equations are ten in number, and contain the ten independent metric components $g_{\mu\nu}$. But we also have the four additional equations

$$G^{\mu\nu}{}_{,\nu} = 0 \qquad \textbf{(3.2.4)}$$

imposed by the Bianchi identities, implying that only six of the ten Einstein equations are independent. This means that we will have ten unknown variables, but only six equations, giving us the freedom to impose four more conditions. We will use this freedom in an attempt to find a coordinate system (specifying a coordinate system in space-time consumes 4 degrees of freedom), in which the four conditions (3.2.1) are satisfied. So, the question is now how to find such a coordinate system. In our quest of finding the appropriate coordinates, we will use a coordinate transformation called the *gauge transformation*.

The gauge transformation

A transformation from the coordinates $\{x^{\alpha}\}$ to the coordinates $\{x^{\alpha'}\}$ of the form

$$x^{\alpha'} = x^{\alpha} + \xi^{\alpha}(x^{\beta}) \qquad \textbf{(3.2.5)}$$

where $|\xi^{\alpha}{}_{,\beta}| << 1$, is called a gauge transformation. And the corresponding transformation matrix is given by

$$\Lambda^{\alpha'}_{\beta} = \delta^{\alpha}_{\beta} + \xi^{\alpha}{}_{,\beta} \qquad \textbf{(3.2.6)}$$

The matrix of the inverse transformation $\Lambda^{\alpha}_{\beta'}$ can be derived in the first order approximation by assuming it has the form $\Lambda^{\alpha}_{\beta'} = \delta^{\alpha}_{\beta} + A^{\alpha}_{\beta}$, where $|A^{\alpha}_{\beta}| << 1$, and using the condition $\delta^{\alpha}_{\sigma} = \Lambda^{\beta'}_{\sigma}\Lambda^{\alpha}_{\beta'}$, i.e.

$$\delta^{\alpha}_{\sigma} = (\delta^{\beta}_{\sigma} + \xi^{\beta}{}_{,\sigma})(\delta^{\alpha}_{\beta} + A^{\alpha}_{\beta}) \approx \delta^{\beta}_{\sigma}\delta^{\alpha}_{\beta} + \xi^{\beta}{}_{,\sigma}\delta^{\alpha}_{\beta} + \delta^{\beta}_{\sigma}A^{\alpha}_{\beta} = \delta^{\alpha}_{\sigma} + \xi^{\alpha}{}_{,\sigma} + A^{\alpha}_{\sigma} \quad \Rightarrow$$

$$A^{\alpha}_{\sigma} = -\xi^{\alpha}{}_{,\sigma} \qquad \textbf{(3.2.7)}$$

This implies that the matrix of the inverse gauge transformation, in the first order approximation is

$$\Lambda^{\alpha}_{\beta'} = \delta^{\alpha}_{\beta} - \xi^{\alpha}{}_{,\beta} \qquad \textbf{(3.2.8)}$$

The metric in a first order approximation after a gauge transformation

The metric in a first order approximation after a gauge transformation is given by

$$g^{(New)}_{\mu\nu} = \Lambda^{\alpha}_{\mu'}\Lambda^{\beta}_{\nu'}g_{\alpha\beta} = (\delta^{\alpha}_{\mu} - \xi^{\alpha}{}_{,\mu})(\delta^{\beta}_{\nu} - \xi^{\beta}{}_{,\nu})(\eta_{\alpha\beta} + h_{\alpha\beta}) \approx$$

$$\approx \delta^{\alpha}_{\mu}\delta^{\beta}_{\nu}\eta_{\alpha\beta} + \delta^{\alpha}_{\mu}\delta^{\beta}_{\nu}h_{\alpha\beta} - \eta_{\alpha\beta}(\delta^{\beta}_{\nu}\xi^{\alpha}{}_{,\mu} + \delta^{\alpha}_{\mu}\xi^{\beta}{}_{,\nu}) \quad \Rightarrow$$

$$g^{(New)}_{\mu\nu} = \eta_{\mu\nu} + h_{\mu\nu} - \xi_{\nu,\mu} - \xi_{\mu,\nu} \qquad \textbf{(3.2.9)}$$

It is now easy to identify the transformed metric perturbation, expressed in the first order approximation

$$h^{(New)}_{\mu\nu} = h_{\mu\nu} - \xi_{\nu,\mu} - \xi_{\mu,\nu} \qquad \textbf{(3.2.10)}$$

The trace reverse of the metric perturbation after a gauge transformation

The transformed trace reverse of the metric perturbation can now with the help of (3.2.10) and (3.1.8) be derived

$$\bar{h}^{(New)}_{\mu\nu} = h^{(New)}_{\mu\nu} - \frac{\eta_{\mu\nu}}{2}h^{(new)} = h_{\mu\nu} - \xi_{\nu,\mu} - \xi_{\mu,\nu} - \frac{\eta_{\mu\nu}}{2}\left(h^{\alpha}{}_{\alpha} - 2\xi^{\alpha}{}_{,\alpha}\right) \quad \Rightarrow$$

$$\bar{h}^{(New)}_{\mu\nu} = \bar{h}_{\mu\nu} - \xi_{\nu,\mu} - \xi_{\mu,\nu} + \eta_{\mu\nu}\xi^{\alpha}{}_{,\alpha} \qquad \textbf{(3.2.11)}$$

If it is now possible to chose the gauge transformation so that $\bar{h}^{(New)\mu\nu}{}_{,\nu} = 0$, the goal of finding an appropriate coordinate system has been achieved. The condition on the gauge transformation that guarantees that $\bar{h}^{(New)\mu\nu}{}_{,\nu} = 0$, can be extracted in the following way

$$\bar{h}^{(new)\mu\nu}{}_{,\nu} = 0 \quad \Leftrightarrow$$

$$\bar{h}^{\mu\nu}{}_{,\nu} - \xi^{\nu,\mu}{}_{,\nu} - \xi^{\mu,\nu}{}_{,\nu} + \eta^{\mu\nu}\xi^{\alpha}{}_{,\alpha\nu} = 0 \quad \Leftrightarrow$$

$$\bar{h}^{\mu\nu}{}_{,\nu} - \xi^{\nu,\mu}{}_{,\nu} - \xi^{\mu,\nu}{}_{,\nu} + \xi^{\alpha}{}_{,\alpha}{}^{,\mu} = 0 \quad \Leftrightarrow$$

$$\Box\xi^{\mu} = \bar{h}^{\mu\nu}{}_{,\nu} \qquad \textbf{(3.2.12)}$$

Since (3.2.12) is the three-dimensional inhomogeneous wave equation, and it always has a solution for any sufficiently well behaved $\bar{h}^{\mu\nu}$ [10], we can always find a gauge transformation that satisfies (3.2.12). The conclusion is thus that one always can find a coordinate system in which the Einstein tensor takes on the form of (3.2.2). This implies that when dealing with weak gravitational fields, the Einstein equations can always be expressed as

$$\Box\bar{h}^{\mu\nu} = -2\kappa T^{\mu\nu} \qquad \textbf{(3.2.13)}$$

This is the three-dimensional inhomogeneous wave equation, which gave birth to the theory of gravitational radiation.

4 Gravitational radiation

In this chapter we will study the properties of gravitational waves and their effect on free particles. We will also investigate how gravitational waves are generated.

In the first two sections of this chapter (sec 4.1-4.2), we will discuss the importance of choosing an appropriate set of coordinates, in which to describe the gravitational waves. We will see that a badly chosen set of coordinates, not only makes it difficult to interpret the solutions of the linearized equations, but can if one is unlucky mimic the appearance of a gravitational wave, even though no real physical wave exists.

In section 4.3 the effect that gravitational waves have on free particles will be investigated.

In the last three sections (sec 4.4-4.6), the solution of the Einstein linear equations will be presented , together with a recipe, describing how this solution can be used to derive the metric perturbations corresponding to gravitational waves emitted by a neutron star binary.

4.1 Gravitational waves in empty space

The linearized Einstein equations for weak gravitational fields in empty space ($T^{\mu\nu}=0$) are

$$\Box \bar{h}^{\mu\nu}=0 \tag{4.1.1}$$

Since these equations simply are homogeneous wave equations, they suggest a plane wave solution of the form

$$h^{\mu\nu}=\mathrm{Re}\left(A^{\mu\nu}e^{ik_\alpha x^\alpha}\right) \tag{4.1.2}$$

where $A^{\mu\nu}$ are constants representing the amplitude of the wave and k_α the propagation vector. This solution implies that

$$h^{\mu\nu}{}_{,\alpha\beta}=-k_\alpha k_\beta h^{\mu\nu} \tag{4.1.3}$$

The relation (4.1.3) together with (3.1.4) and (3.1.5) reveals the following identities for the Riemann- and the Ricci-tensor

$$R^\alpha{}_{\mu\beta\nu}=\frac{1}{2}\left(k_\mu k_\beta h^\alpha{}_\nu+k^\alpha k_\nu h_{\beta\mu}-k_\mu k_\nu h^\alpha{}_\beta-k^\alpha k_\beta h_{\mu\nu}\right) \tag{4.1.4}$$

$$R_{\mu\nu}=\frac{1}{2}\left(k_\mu k_\alpha h^\alpha{}_\nu+k_\alpha k_\nu h^\alpha{}_\mu-k_\mu k_\nu h^\alpha{}_\alpha-k^\alpha k_\alpha h_{\mu\nu}\right) \tag{4.1.5}$$

By making the substitution

$$w_\nu=k_\alpha h^\alpha{}_\nu-\frac{1}{2}k_\nu h^\alpha{}_\alpha \tag{4.1.6}$$

the expression (4.1.5) for the Ricci tensor can be rewritten as

$$R_{\mu\nu}=\frac{1}{2}\left(k_\mu w_\nu+k_\nu w_\mu-k^\alpha k_\alpha h_{\mu\nu}\right) \tag{4.1.7}$$

Now since the non-linearized Einstein equations can be rewritten in the form

$$R_{\mu\nu} = \kappa\left(T_{\mu\nu} - \frac{g_{\mu\nu}}{2}T\right) \tag{4.1.8}$$

the non-linearized Einstein equations in empty space can be expressed as

$$R_{\mu\nu} = 0 \tag{4.1.9}$$

Then (4.1.7) together with (4.1.9) imply

$$k^{\alpha}k_{\alpha}h_{\mu\nu} = k_{\mu}w_{\nu} + k_{\nu}w_{\mu} \tag{4.1.10}$$

There are two interesting cases in which we will study the above relation.

a) $k^{\alpha}k_{\alpha} \neq 0$

In this case we get that $$h_{\mu\nu} = \frac{1}{k^{\alpha}k_{\alpha}}\left(k_{\mu}w_{\nu} + k_{\nu}w_{\mu}\right) \tag{4.1.11}$$

Substituting this in equation (4.1.4) we get

$$R^{\alpha}{}_{\mu\beta\nu} = \frac{1}{2k^{\alpha}k_{\alpha}}\left[k_{\mu}k_{\beta}(k^{\alpha}w_{\nu} + k_{\nu}w^{\alpha}) + k^{\alpha}k_{\nu}(k_{\beta}w_{\mu} + k_{\mu}w_{\beta}) - k_{\mu}k_{\nu}(k^{\alpha}w_{\beta} + k_{\beta}w^{\alpha}) - \right.$$

$$\left. - k^{\alpha}k_{\beta}(k_{\mu}w_{\nu} + k_{\nu}w_{\mu})\right] = 0$$

This means that space-time is flat, and the "wave" is just a product of a poorly chosen coordinate system.

b) $k^{\alpha}k_{\alpha} = 0$

This implies that $$k_{\mu}w_{\nu} + k_{\nu}w_{\mu} = 0 \tag{4.1.12}$$

If we assume that the wave travels in the x^3*-direction* of the coordinate system, then $k^{\alpha}k_{\alpha} = 0$ implies that

$$k_{\mu} = (k_0, 0, 0, k_0) \tag{4.1.13}$$

Now equation (4.1.12) and (4.1.13) reveal the following

$$(\delta_\mu^0 + \delta_\mu^3)w_\nu + (\delta_\nu^0 + \delta_\nu^3)w_\mu = 0 \quad \Rightarrow$$

$$w_\nu = 0 \quad \Leftrightarrow$$

$$k_\alpha h^\alpha{}_\nu = \frac{1}{2} k_\nu h^\alpha{}_\alpha \qquad \textbf{(4.1.14)}$$

This relation is called a *gauge condition*, and it will not make the Riemann tensor identical with zero. This means that the only physically significant solutions to Einstein's linear equations must satisfy the condition $k^\alpha k_\alpha = 0$, i.e., the propagation vector of the gravitational wave is a null vector, which also implies that gravity waves travel with the speed of light. Another nice thing about this condition is that it provides us with the gauge condition (4.1.14). In the next section, this condition will helps us to choose a coordinate system in which the solutions to (4.1.1) are easily interpreted.

4.2 The TT-gauge

In this section we will use the gauge condition (4.1.14) to express the plane wave solution of (4.1.1), in a way that makes it simple for us to extract the physical properties of the wave.

We have already seen that the necessary condition $k^\alpha k_\alpha = 0$ implies the gauge condition (4.1.14). This condition is surely satisfied if $h^\alpha{}_\alpha = 0$ and $k_\alpha h^{\alpha\nu} = 0$, which is equivalent (see (3.1.8) & (3.1.9)) to the conditions

$$\bar{h}^\alpha{}_\alpha = 0 \qquad \textbf{(4.2.1)}$$

$$k_\alpha \bar{h}^{\alpha\nu} = 0 \qquad \textbf{(4.2.2)}$$

However if we assume that the wave travels in the x^3 –direction, it is easy to see that (4.2.2) is satisfied by

$$\bar{h}^{0\mu} = 0 \qquad \textbf{(4.2.3)}$$

$$\bar{h}^{3\mu} = 0 \qquad \textbf{(4.2.4)}$$

Let us assume that we already have chosen a coordinate system such that the Lorentz gauge condition $\bar{h}^{\mu\nu}{}_{,\nu} = 0$ is satisfied, which means that the Einstein equations can be given the form of (4.1.1) and the solution is a plane wave

$$\bar{h}^{\mu\nu} = \mathrm{Re}\left(C^{\mu\nu} e^{ik_\alpha x^\alpha}\right) \qquad \textbf{(4.2.5)}$$

We now want to explore the possibility of finding a new coordinate system, which is such that both the Lorentz gauge condition (which is equivalent to (4.2.2)) and the additional conditions (4.2.1), (4.2.3) and (4.2.4) are satisfied.

To find the desired coordinate system we will make a gauge transformation. This transformation must be such that it preserves the Lorentz gauge, and thus must satisfy

$$\Box \xi^{\mu} = \bar{h}^{\mu\nu}{}_{,\nu} = 0 \tag{4.2.6}$$

This means that the gauge transformation will have the form of a plane wave

$$\xi_{\mu} = B_{\mu} e^{ik_{\alpha}x^{\alpha}}, \quad (|B_{\mu}| << 1) \tag{4.2.7}$$

where B_{μ} are constants.
We already know from equation (3.2.11) that trace reverse of the metric perturbation after a gauge transformation is expressed by

$$\bar{h}^{(New)}_{\mu\nu} = \mathrm{Re}\left(C^{(New)}_{\mu\nu} e^{ik_{\alpha}x^{\alpha}}\right) = \bar{h}_{\mu\nu} - \xi_{\nu,\mu} - \xi_{\mu,\nu} + \eta_{\mu\nu}\xi^{\alpha}{}_{,\alpha} \tag{4.2.8}$$

This relation together with (4.2.5) and (4.2.7) imply

$$C^{(New)}_{\mu\nu} = C_{\mu\nu} - ik_{\mu}B_{\nu} - ik_{\nu}B_{\mu} + i\eta_{\mu\nu}B^{\alpha}k_{\alpha} \tag{4.2.9}$$

Now we want to know if it is possible to choose B_{μ} in such a way that (4.2.1) and (4.2.3) are satisfied by $\bar{h}^{(New)}_{\mu\nu}$, i.e., we want to know if we can find coefficients B_{μ} such that

$$C^{(New)\nu}{}_{\nu} = 0 \tag{4.2.10}$$

$$C^{(New)}_{0\mu} = 0 \tag{4.2.11}$$

(Here we have omitted the condition (4.2.4) since the gauge transformation preserves the Lorentz gauge, and thus automatically satisfies (4.2.4) whenever (4.2.3) is true)

We start our search for the desired coefficients by raising the index μ and letting it be equal to ν in (4.2.9), then by using (4.2.10) we get that

$$B^{\nu}k_{\nu} = \frac{i}{2}C^{\nu}{}_{\nu} \tag{4.2.12}$$

Then by putting $\mu = 0$ in (4.2.9) together with the use of (4.2.11) and (4.2.12) it is not too difficult to show that

$$B_0 = -\frac{i}{2k_0}(C_{00} - \frac{1}{2}C^{\nu}{}_{\nu}) \tag{4.2.13}$$

Finally by letting $\nu = i$ (Roman letters denote spatial coordinates) in (4.2.9) and using (4.2.13) it is easy to derive the following identity

$$B_i = \frac{i}{2(k_0)^2}\left[-2k_0 C_{0i} + k_i\left(C_{00} - \frac{1}{2}C^{\nu}{}_{\nu}\right)\right] \tag{4.2.14}$$

The existence of the identities (4.2.13) and (4.2.14) tells us that it is always possible to find a coordinate system in which the conditions (4.2.1) and (4.2.3) together with the Lorentz gauge condition are satisfied. These three conditions together are known as the *transverse-traceless* (TT) gauge conditions. In the TT-gauge the metric perturbation of a gravitational wave travelling in the x^3-direction has the form

$$h_{\mu\nu}^{TT} = \begin{pmatrix} 0 & 0 & 0 & 0 \\ 0 & C_+ & C_\times & 0 \\ 0 & C_\times & -C_+ & 0 \\ 0 & 0 & 0 & 0 \end{pmatrix} \cos(k_0 ct - k_0 x^3) \tag{4.2.15}$$

where $k_0 c = \omega$ is the frequency of the wave.

4.3 The polarization of gravitational waves and their effect on free particles

Let us now continue to investigate what happens when a gravitational wave interacts with free particles, and with free particles we mean that there are no forces acting on the particles.

To do so, we will consider the case when two free particles in the *xy*-plane, separated by the distance ξ^α, are "hit" by an incoming gravitational wave travelling in the *z*-direction (the x^3-direction). We now know that the distance between two geodesics that are parallel at some point, is described by the equation of geodesic deviation, and since the geodesics of the two particles are parallel before the wave "hits" the *xy*-plane, we can use this equation to determine how the distance between the particles evolves in time.
By assuming that the particles' velocities are much smaller than the speed of light, and by using (1.12.10) we get the following approximate relation

$$\frac{D^2\xi^\alpha}{dt^2} = c^2 R^\alpha{}_{00\nu}\xi^\nu \tag{4.3.1}$$

Since the gravitational field associated with the wave is weak, we can assume the good approximation where the covariant derivative can be replaced by the conventional partial derivative. This means that we can rewrite equation (4.3.1) to

$$\frac{\partial^2\xi^\alpha}{\partial t^2} = c^2 R^\alpha{}_{00\nu}\xi^\nu \tag{4.3.2}$$

We then continue with the calculation of the components of the Riemann tensor that appear in (4.3.2), by using (4.2.15) and (3.1.4)

$$R^{\alpha}{}_{00\nu} = \frac{\eta^{\alpha\sigma}}{2} h^{TT}_{\sigma\nu,00} = \frac{\eta^{\alpha\sigma}}{2c^2} \frac{\partial^2 h^{TT}_{\sigma\nu}}{\partial t^2} \quad \textbf{(4.3.3)}$$

Finally, putting (4.3.2) and (4.3.2) together we arrive at the equations that in the first order approximation describe how the distance between the two particles evolves in time when they are being "hit" by a gravitational wave

$$\frac{\partial^2 \xi^{\alpha}}{\partial t^2} = \frac{\eta^{\alpha\sigma}}{2} \frac{\partial^2 h^{TT}_{\sigma\nu}}{\partial t^2} \xi^{\nu} \quad \textbf{(4.3.4)}$$

Since the only non-zero components of the metric perturbation are given by $h^{TT}_{11} = -h^{TT}_{22} = C_{+} \cos(\omega t - k_0 z)$ and $h^{TT}_{12} = h^{TT}_{21} = C_{\times} \cos(\omega t - k_0 z)$, the equations (4.3.4) are equivalent to

$$\frac{\partial^2 \xi^1}{\partial t^2} = -\frac{1}{2}\left[\xi^1 \frac{\partial^2}{\partial t^2}(C_{+} \cos(\omega t - k_0 z)) + \xi^2 \frac{\partial^2}{\partial t^2}(C_{\times} \cos(\omega t - k_0 z))\right] \quad \textbf{(4.3.5)}$$

$$\frac{\partial^2 \xi^2}{\partial t^2} = -\frac{1}{2}\left[\xi^1 \frac{\partial^2}{\partial t^2}(C_{\times} \cos(\omega t - k_0 z)) - \xi^2 \frac{\partial^2}{\partial t^2}(C_{+} \cos(\omega t - k_0 z))\right] \quad \textbf{(4.3.6)}$$

The two independent components, C_{+} & $C_{\times}$ of the metric perturbation, correspond to the two polarization directions of the gravitational wave. We will now continue to explore how these directions manifest themselves by investigating what happens when gravitational waves of different polarization "hit" a ring of free test particles in the $z = 0$ plane.

C_{+}-Polarized wave

In this case we have $C_{\times} = 0$, this means that equation (4.3.5) transforms to

$$\frac{\partial^2 \xi^1}{\partial t^2} = -\frac{1}{2} \xi^1 \frac{\partial^2}{\partial t^2}(C_{+} \cos(\omega t - k_0 z)) \quad \textbf{(4.3.7)}$$

The solution of this equation can be found by assuming that the solution has the form

$$\xi^1(t) = \xi^1(0) + \sigma^1(t) \quad \textbf{(4.3.8)}$$

where $\xi^1(0)$ is the distance between two test particles before the wave reaches the $z = 0$ plane, and where $\sigma(t)$ represents the change in length that the distance undergoes when the wave "hits" the plane, having the property $|\sigma^1(t)| << \xi^1(0)$. Substituting ξ^1 on the right-hand side of equation (4.3.7) with (4.3.8) we get in the first order approximation (omitting all powers of C_{+} & $\sigma^1(t)$ greater than one)

$$\frac{\partial^2 \xi^1}{\partial t^2} = -\frac{1}{2} \xi^1(0) \frac{\partial^2}{\partial t^2}(C_{+} \cos(\omega t - k_0 z)) \quad \textbf{(4.3.9)}$$

The solution to this equation is

$$\xi^1 = B + At - \xi^1(0)\frac{C_+}{2}\cos(\omega t - k_0 z) \tag{4.3.10}$$

where A and B are constants. Comparing this solution to (4.3.8), the following identifications can be made

$$B = \xi^1(0) \tag{4.3.11}$$

$$\sigma^1(t) = At - \xi^1(0)\frac{C_+}{2}\cos(\omega t - k_0 z) \tag{4.3.12}$$

We also remind ourselves of the condition $|\sigma^1(t)| << \xi^1(0)$ which implies that $A=0$. The approximate solution to (4.3.7) is then

$$\xi^1 = \xi^1(0)\left[1 - \frac{C_+}{2}\cos(\omega t - k_0 z)\right] \tag{4.3.13}$$

By letting $C_\times = 0$ in equation (4.3.6) and by using the exact same method that were used in to find the solution (4.3.13) we get

$$\xi^2 = \xi^2(0)\left[1 + \frac{C_+}{2}\cos(\omega t - k_0 t)\right] \tag{4.3.14}$$

Since the test particles are placed in a circle, the initial separation between two particles that are separated by a line going through the centre of the circle can be expressed as

$$(\xi^1(0), \xi^2(0)) = (d\cos(\theta), d\sin(\theta)) \tag{4.3.15}$$

where d is the diameter of the circle and θ is the angle between the separation line and the x^1 –axis. By using (4.3.14) and (4.3.15) the positions of the particles when the wave has reached the $z=0$ -plane can be derived

$$x^1 = \frac{\xi^1}{2} = a(t)\cos(\theta) \quad ; \quad x^2 = \frac{\xi^2}{2} = b(t)\sin(\theta) \tag{4.3.16}$$

where

$$a(t) = \frac{d}{2}\left[1 - \frac{C_+}{2}\cos(\omega t)\right] \tag{4.3.17}$$

$$b(t) = \frac{d}{2}\left[1 + \frac{C_+}{2}\cos(\omega t)\right] \tag{4.3.18}$$

Now three interesting cases occur

a) $t = \frac{\pi}{2\omega} n \;\; ; \;\; n \in \boldsymbol{N}$

In this case $a = b = \frac{d}{2}$,and the particles will be placed in there initial positions of the circle.

b) $t = \frac{2\pi}{\omega} n \;\; ; \;\; n \in \boldsymbol{N}$

In this case $a = \frac{d}{2}\left[1 - \frac{C_+}{2}\right]$, $b = \frac{d}{2}\left[1 + \frac{C_+}{2}\right]$, and the particles will be placed along the boundary of an ellipse with its principal axis parallel to the x^2-axis.

c) $t = \frac{\pi}{\omega}(2n+1) \;\; ; \;\; n \in \boldsymbol{N}$

In this case $a = \frac{d}{2}\left[1 + \frac{C_+}{2}\right]$, $b = \frac{d}{2}\left[1 - \frac{C_+}{2}\right]$, and the particles will be placed along the boundary of an ellipse with its principal axis parallel to the x^1*-axis*.

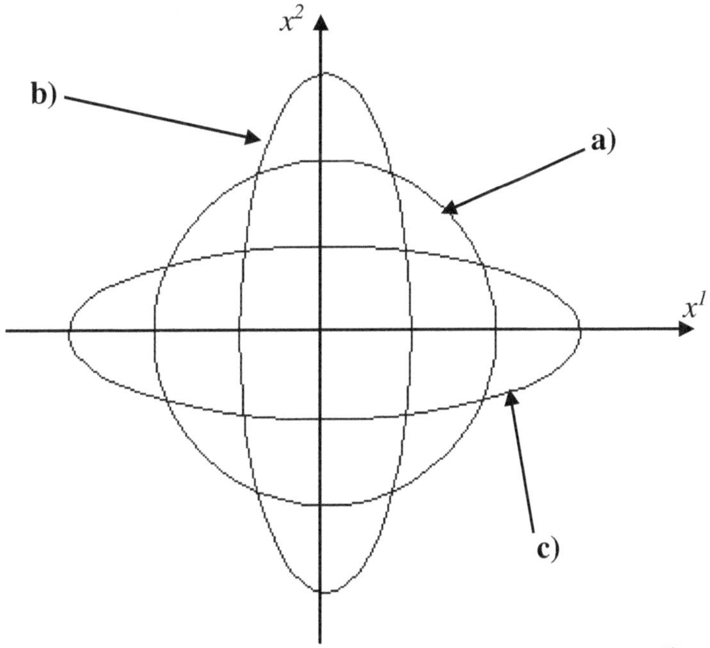

(figure 6)

We now see that when a C_+-polarized wave travelling in the z-direction "hits" the plane with the test particles, the particles will be "moved" in the cross-like pattern of the figure above.

$C_\times$-polarized wave

In this case we have $C_+ = 0$, and consequently the equations (4.3.5) & (4.3.6) take on the form

$$\frac{\partial^2}{\partial t^2}\begin{pmatrix}\xi^1\\ \xi^2\end{pmatrix}=\begin{pmatrix}0 & -\frac{\partial^2}{\partial t^2}\left(\frac{C_\times}{2}\cos(\omega t-k_0 z)\right)\\ -\frac{\partial^2}{\partial t^2}\left(\frac{C_\times}{2}\cos(\omega t-k_0 z)\right) & 0\end{pmatrix}\begin{pmatrix}\xi^1\\ \xi^2\end{pmatrix} \quad \textbf{(4.3.19)}$$

By assuming that the solution has the form

$$\xi^1(t)=\xi^1(0)+\sigma^1(t) \quad ; \quad \xi^2(t)=\xi^2(0)+\sigma^2(t) \quad \textbf{(4.3.20)}$$

where $|\sigma^1(t)| << \xi^1(0)$ & $|\sigma^2(t)| << \xi^2(0)$, we can use the same method to solve (4.3.19) as we used to derive the solution of (4.3.7). This gives the solution

$$\begin{pmatrix}\xi^1\\ \xi^2\end{pmatrix}=\begin{pmatrix}1 & -\frac{C_\times}{2}\cos(\omega t-k_0 z)\\ -\frac{C_\times}{2}\cos(\omega t-k_0 z) & 1\end{pmatrix}\begin{pmatrix}\xi^1(0)\\ \xi^2(0)\end{pmatrix} \quad \textbf{(4.3.21)}$$

In order to make it easier to interpret this solution, we make a coordinate transformation by changing coordinates to a coordinate system that is rotated 45 degrees relative to the original coordinate system. If we denote the 2×2 matrix appearing in (4.3.21) by $\overline{\overline{M}}$, we get that the solution (4.3.21) in the new coordinate system is given by

$$\begin{pmatrix}\xi^{1'}\\ \xi^{2'}\end{pmatrix}=\begin{pmatrix}\cos(\pi/4) & \sin(\pi/4)\\ -\sin(\pi/4) & \cos(\pi/4)\end{pmatrix}\overline{\overline{M}}\begin{pmatrix}\cos(\pi/4) & -\sin(\pi/4)\\ \sin(\pi/4) & \cos(\pi/4)\end{pmatrix}\begin{pmatrix}\xi^{1'}(0)\\ \xi^{2'}(0)\end{pmatrix}$$

$$\Leftrightarrow$$

$$\begin{pmatrix}\xi^{1'}\\ \xi^{2'}\end{pmatrix}=\begin{pmatrix}1-\frac{C_\times}{2}\cos(\omega t-k_0 z) & 0\\ 0 & 1+\frac{C_\times}{2}\cos(\omega t-k_0 z)\end{pmatrix}\begin{pmatrix}\xi^{1'}(0)\\ \xi^{2'}(0)\end{pmatrix} \quad \textbf{(4.3.22)}$$

Comparing (4.3.22) with the solutions (4.3.13) and (4.3.14), it is clear that when the gravitational wave is $C_\times$-polarized, the test particles will "move" in a pattern that is similar to the pattern that occurs when the wave is C_+-polarized. The only difference is that the two patterns are rotated 45 degrees relative to each other, see figure 7.

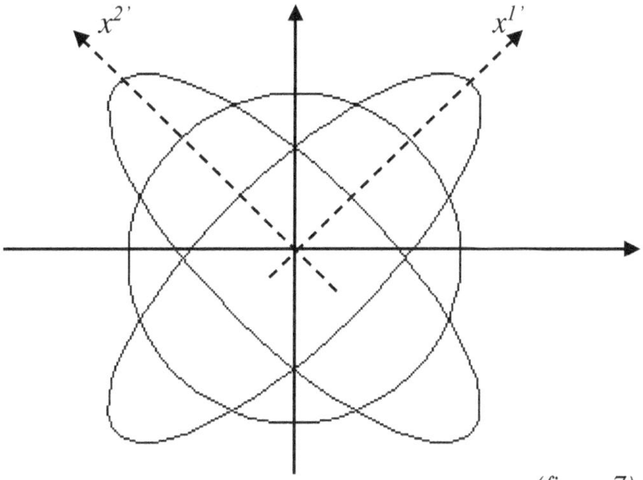

(figure 7)

Circularly polarized wave

A circularly polarized gravitational wave is the composition of one C_+-polarized and one $C_\times$-polarized wave. These two waves have the same amplitude ($C_+ = C_\times = C_o$) and are 90 degrees out of phase with each other. This means that the equations (4.3.5) and (4.3.6) take on the form

$$\frac{\partial^2}{\partial t^2}\begin{pmatrix} \xi^1 \\ \xi^2 \end{pmatrix} = \frac{\partial^2}{\partial t^2}\begin{pmatrix} -\frac{C_o}{2}\cos(\omega t) & \frac{C_o}{2}\sin(\omega t) \\ \frac{C_o}{2}\sin(\omega t) & \frac{C_o}{2}\cos(\omega t) \end{pmatrix}\begin{pmatrix} \xi^1(0) \\ \xi^2(0) \end{pmatrix} \tag{4.3.23}$$

Here we have assumed that the wave has reached the *z=0* –plane.

By using the same methods that were used in the C_+-polarized and the $C_\times$-polarized cases, the solution to (4.3.23) is easily derived

$$\begin{pmatrix} \xi^1 \\ \xi^2 \end{pmatrix} = \begin{pmatrix} 1-\frac{C_0}{2}\cos(\omega t) & \frac{C_o}{2}\sin(\omega t) \\ \frac{C_o}{2}\sin(\omega t) & 1+\frac{C_0}{2}\cos(\omega t) \end{pmatrix}\begin{pmatrix} \xi^1(0) \\ \xi^2(0) \end{pmatrix} \tag{4.3.24}$$

In order to make it easier to interpret this solution, we make a coordinate transformation by changing coordinates to a coordinate system, which is rotated at an angle of $-\omega t/2$ radians relative to the original coordinate system. If we denote the 2×2 matrix appearing in (4.3.24) by $\overline{\overline{M}}$, we get that the solution (4.3.21) in the new coordinate system is given by

$$\begin{pmatrix} \xi^{1'} \\ \xi^{2'} \end{pmatrix} = \begin{pmatrix} \cos(\omega t/2) & -\sin(\omega t/2) \\ \sin(\omega t/2) & \cos(\omega t/2) \end{pmatrix}\overline{\overline{M}}\begin{pmatrix} \cos(\omega t/2) & \sin(\omega t/2) \\ -\sin(\omega t/2) & \cos(\omega t/2) \end{pmatrix}\begin{pmatrix} \xi^{1'}(0) \\ \xi^{2'}(0) \end{pmatrix}$$

$$\Leftrightarrow$$

$$\begin{pmatrix} \xi^{1'} \\ \\ \xi^{2'} \end{pmatrix} = \begin{pmatrix} 1-\frac{C_o}{2} & 0 \\ 0 & 1+\frac{C_o}{2} \end{pmatrix} \begin{pmatrix} \xi^{1'}(0) \\ \\ \xi^{2'}(0) \end{pmatrix} \qquad \textbf{(4.3.25)}$$

This means that if the test particles' initial positions are on the perimeter of a circle of radius r, their positions after the circularly polarized wave have "hit" the $z=0$ –plane, will be given by

$$x^{1'} = \frac{\xi^{1'}}{2} = \left[1-\frac{C_o}{2}\right]\cos(\theta) \quad ; \quad x^{2'} = \frac{\xi^{2'}}{2} = \left[1+\frac{C_o}{2}\right]\sin(\theta) \qquad \textbf{(4.3.26)}$$

where $0 \le \theta \le 2\pi$.

In other words, the positions of the test particles will be at the perimeter of an ellipse with its principal axis parallel to the $x^{2'}$-axis, and since the rotation angle of the primed coordinate system is given by $-\omega t/2$, the ellipse will rotate in a way given by the figure below

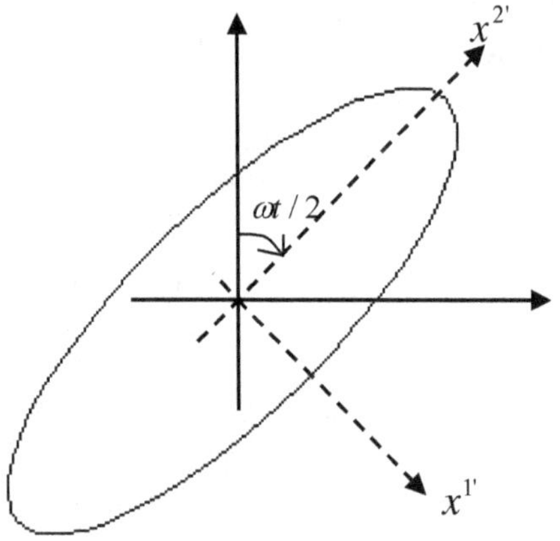

(figure 8)

Observe that the individual test-particles do not move around the origin of the coordinate system; they just move in little epicycles[11].

4.4 Solution to Einstein's linearized field equations

The equation we will solve here is the linearized field equations

$$\left(\frac{\partial^2}{\partial t^2} - \nabla^2\right)\bar{h}^{\mu\nu}(\bar{r},t) = -2\kappa T^{\mu\nu}(\bar{r},t) \qquad \textbf{(4.4.1)}$$

This is the same inhomogeneous wave equation that appears in electromagnetic wave theory and the well known solution to this equation is given by the retarded integral [12]

$$\bar{h}^{\mu\nu}(\bar{r},t) = -\frac{\kappa}{2\pi}\int_{V'} \frac{T^{\mu\nu}(\bar{r}',t-R/c)}{R}dV' \qquad \textbf{(4.4.2)}$$

where $R = |\bar{r} - \bar{r}'|$.

We will assume that $|\bar{r}| >> |\bar{r}'|$, i.e. we calculate the metric perturbation far away from the source. This means that $R \approx |\bar{r}|$, which enables us to make the following approximation of the solution

$$\bar{h}^{\mu\nu}(\bar{r},t) = -\frac{\kappa}{2\pi r}\int_{V'} T^{\mu\nu}(\bar{r}',t-r/c)dV' \qquad \textbf{(4.4.3)}$$

It is now possible to simplify the solution (4.4.3) considerably by first making use of the fact that $T^{\mu\nu}{}_{,\nu} = 0$ in the weak field approximation, i.e.

$$T^{\mu 0}{}_{,0} = -T^{\mu k}{}_{,k} \qquad \textbf{(4.4.4)}$$

where the roman letter *k* represents spatial coordinates. This relation together with gauss theorem imply that

$$\bar{h}^{\mu 0}{}_{,0} = -\frac{\kappa}{2\pi r}\int_{V'} T^{\mu 0}{}_{,0}dV' = \frac{\kappa}{2\pi r}\int_{V'} T^{\mu k}{}_{,k}dV' = \frac{\kappa}{2\pi r}\int_{S'} T^{\mu k}dS' \qquad \textbf{(4.4.5)}$$

But since the energy momentum tensor of the source only is non-zero inside a limited region V' of space, it must be zero on the boundary S' of this region, i.e.

$$T^{\mu\nu}(\bar{r}',t-r/c)\Big|_{\bar{r}'\in S'} = 0 \qquad \textbf{(4.4.6)}$$

This observation and (4.4.5) then imply

$$\bar{h}^{\mu 0}{}_{,0} = 0 \qquad \textbf{(4.4.7)}$$

Which means that the only time dependent components of the trace reverse of the metric perturbation are the spatial components

$$\bar{h}^{ik}(\bar{r},t) = -\frac{\kappa}{2\pi r}\int_{V'} T^{ik}(\bar{r},t-r/c)dV' \qquad \textbf{(4.4.8)}$$

The next step of our simplification of the solution (4.4.3) is to prove the relation

$$\int_{V'} T^{ik}(\bar{r}',t-r/c)dV' = \frac{1}{2c^2}\frac{\partial^2}{\partial t^2}\int_{V'} T^{00}(\bar{r}',t-r/c)x^{i'}x^{k'}dV' \qquad \textbf{(4.4.9)}$$

proof

a) First we use (4.4.4), partial integration and (4.4.6) to show

$$\int_V T^{00}{}_{,0}x^i x^k dV = -\int_V T^{0l}{}_{,l}x^i x^k dV = \left[-T^{0l}x^i x^k\right]_{\bar{r}=0}^{\bar{r}\in S} + \int_V T^{0l}\left(\delta^i_l x^k + x^i \delta^k_l\right)dV \Rightarrow$$

$$\int_V T^{00}{}_{,0}x^i x^k dV = \int_V \left(T^{0i}x^k + T^{0k}x^i\right)dV \qquad \textbf{(4.4.10)}$$

b) Then we use (4.4.10), (4.4.4), partial integration and (4.4.6) to show

$$\int_V T^{00}{}_{,00}x^i x^k dV = \int_V \left(T^{i0}{}_{,0}x^k + T^{k0}{}_{,0}x^i\right)dV = -\int_V \left(T^{il}{}_{,l}x^k + T^{kl}{}_{,l}x^i\right)dV =$$

$$= -\left[\left(T^{il}x^k + T^{kl}x^i\right)\right]_{\bar{r}=0}^{\bar{r}\in S} + \int_V \left(T^{il}\delta_l^k + T^{kl}\delta_l^i\right)dV \quad \Rightarrow$$

$$\frac{1}{c^2}\frac{\partial^2}{\partial t^2}\int_V T^{00}x^i x^k dV = 2\int_V T^{ik} dV \qquad \text{which completes the proof.}$$

This implies that the solution to Einstein's linearized equations can be expressed as

$$\bar{h}^{ik}(\bar{r},t) = -\frac{2G}{c^6 r}\frac{\partial^2}{\partial t^2}\int_{V'} T^{00}(\bar{r}',t-r/c)x^{i'}x^{k'}dV' \qquad \textbf{(4.4.11)}$$

For slowly moving sources we also have the approximate relation $T^{00} \approx \rho c^2$ (where ρ is the mass density of the source), which means that the solution to Einstein's linearized equations in the first order approximation can be written as

$$\bar{h}^{ik}(\bar{r},t) = -\frac{2G}{c^4 r}\frac{\partial^2 I^{ik}}{\partial t^2} \qquad \textbf{(4.4.12)}$$

where

$$I^{ik} = \int_{V'} \rho(\bar{r}',t-r/c)x^{i'}x^{k'}dV' \qquad \textbf{(4.4.13)}$$

The tensor I^{ik} is often referred to as the *quadrupole moment tensor of the mass distribution,* or just the *quadrupole moment tensor*. It is important not to forget that (4.4.12-13) is a approximate solution developed within the frame work of weak gravitational fields, only correct to the first order in $\bar{h}^{ik}$. This solution is commonly known as the quadrupole approximation.

4.5 The solution of Einstein's linearized equations expressed in the TT-gauge

We will now find a way to express the solution of the Einstein linear equations in the TT-gauge. In the TT-gauge the trace reverse of the metric perturbation satisfies

$$\bar{h}^{TT\,\alpha}{}_{\alpha} = 0 \quad ; \quad \bar{h}^{TT\,0\mu} = 0 \tag{4.5.1}$$

In order to simplify the search for a suitable gauge transformation, which will help us to express the solutions in the TT-gauge, we begin by observing that the time and spatial dependence on the quadrupole moment tensor, always can be reduced to sums of terms having the form

$$I^{ik} = \mathrm{Re}\left(\tilde{I}^{ik} e^{ik_\alpha x^\alpha}\right) \tag{4.5.2}$$

by Fourier analysis. Where $\tilde{I}^{ik}$ is the constant part of the quadrupole tensor.

The solution (4.4.12) can then be expressed by sums of terms having the form

$$\bar{h}^{ik} = \mathrm{Re}\left(\frac{2G}{c^4 r}\omega^2 \tilde{I}^{ik} e^{ik_\alpha x^\alpha}\right) \tag{4.5.3}$$

where $\omega = k_0 c$ is the frequency of the gravitational wave. This means that in the TT-gauge the solution can be expressed as

$$\bar{h}^{TT\,ik} = \mathrm{Re}\left(\frac{2G}{c^4 r}\omega^2 \tilde{I}^{TT\,ik} e^{ik_\alpha x^\alpha}\right) \tag{4.5.4}$$

where $\tilde{I}^{TT\,ik}$ is the constant part of the quadrupole moment tensor in the TT-gauge

$$I^{TT\,ik} = \mathrm{Re}\left(\tilde{I}^{TT\,ik} e^{ik_\alpha x^\alpha}\right) \tag{4.5.6}$$

We will now continue with the introduction of a gauge transformation defined by

$$\xi^\alpha = \frac{2G}{c^4 r}\omega^2 B^\alpha e^{ik_\alpha x^\alpha} \quad ; \quad |B^\alpha| << 1 \tag{4.5.7}$$

By assuming that the distance r to the source is great we can derive $\xi^\alpha{}_{,\beta}$ and $\xi^{\alpha,\beta}{}_{,\beta}$ to the first order, i.e., omitting the terms containing the factors $1/r^2$ and $1/r^3$.

$$\xi^\alpha{}_{,\beta} \approx k_\beta \xi^\alpha \tag{4.5.8}$$

$$\Box \xi^\alpha = \xi^{\alpha,\beta}{}_{,\beta} \approx k_\beta k^\beta \xi^\alpha \tag{4.5.9}$$

It is now clear from (4.5.9) that if we are dealing with physical waves we will have $\Box \xi^\alpha = 0$, since these waves always have propagation vectors satisfying $k_\alpha k^\alpha = 0$. This implies that the gauge transformation defined by (4.5.7) to the first order preserves the Lorentz gauge.

Let us now assume that the gauge transformation defined above takes us into the TT-gauge, we get from (3.2.10) that

$$\bar{h}_{\mu\nu}^{TT} = \mathrm{Re}\left(\frac{2G}{c^4 r}\omega^2 \tilde{I}_{\mu\nu}^{TT} e^{ik_\alpha x^\alpha}\right) = \bar{h}_{\mu\nu} - \xi_{\nu,\mu} - \xi_{\mu,\nu} + \eta_{\mu\nu}\xi^\alpha{}_{,\alpha} \quad \textbf{(4.5.10)}$$

The expression above together with (4.5.3), (4.5.7) and (4.5.8) then imply

$$\tilde{I}_{\mu\nu}^{TT} = \tilde{I}_{\mu\nu} - ik_\mu B_\nu - ik_\nu B_\mu + i\eta_{\mu\nu}B^\alpha k_\alpha \quad \textbf{(4.5.11)}$$

Since we are now in the TT-gauge the conditions (4.5.1) must be satisfied, thus implying

$$\tilde{I}^{TT\,\nu}{}_\nu = 0 \quad ; \quad \tilde{I}^{TT\,0\mu} = 0 \quad \textbf{(4.5.12)}$$

Comparing (4.5.11) and (4.5.12) with (4.2.9), (4.2.10) and (4.2.11), we see that these expressions are equivalent with the only difference that here $C_{\mu\nu}^{(New)}$ and $C_{\mu\nu}$ have been replaced by $\tilde{I}_{\mu\nu}^{TT}$ and $\tilde{I}_{\mu\nu}$. This means that the coefficients B^α that induce the TT-gauge to the solution of the Einstein linearized equations are given by (4.2.13) and (4.2.14), i.e.

$$B_0 = -\frac{i}{2k_0}\left(\tilde{I}_{00} - \frac{1}{2}\tilde{I}^\nu{}_\nu\right) \quad \textbf{(4.5.13)}$$

$$B_i = \frac{i}{2(k_0)^2}\left[-2k_0\tilde{I}_{0i} + k_i(\tilde{I}_{00} - \frac{1}{2}\tilde{I}^\nu{}_\nu)\right] \quad \textbf{(4.5.14)}$$

Since the only time dependent components of the solution to the linearized Einstein equations are the spatial components (se (4.4.7)), we have that

$$\tilde{I}^{0\mu} = 0 \quad \textbf{(4.5.15)}$$

If we also assume that the gravitational wave travels in the *z*-direction (i.e. $k_\alpha = (k_0, 00, k_0)$), the Lorentz gauge condition ($\bar{h}^{\mu\nu}{}_{,\nu} = 0$) together with (4.5.15) imply that

$$\tilde{I}^{3\mu} = 0 \quad \textbf{(4.5.16)}$$

Then the conditions (4.5.15) and (4.5.16) simplify the expressions for the coefficients B_α in the following way

$$B_0 = -\frac{i}{4k_0}(\tilde{I}_{xx} + \tilde{I}_{yy}) \quad \textbf{(4.5.17)}$$

$$B_i = \frac{i}{4k_0}(\tilde{I}_{xx} + \tilde{I}_{yy})\delta_i^z \quad \textbf{(4.5.18)}$$

Finally the two above relations together with (4.5.11) and the assumption that the wave travels in the *z*-direction, imply that the constant coefficients of the quadrupole moment tensor in the TT-gauge are given by

$$\widetilde{I}_{zi}^{TT} = 0 \;\; ; \;\; \widetilde{I}_{xx}^{TT} = -\widetilde{I}_{yy}^{TT} = \frac{1}{2}\left(\widetilde{I}_{xx} - \widetilde{I}_{yy}\right) \;\; ; \;\; \widetilde{I}_{xy}^{TT} = \widetilde{I}_{xy} \qquad \textbf{(4.5.19)}$$

Similarly, the above relations must also be true for the quadrupole moment tensor, i.e.

$$I_{zi}^{TT} = 0 \;\; ; \;\; I_{xx}^{TT} = -I_{yy}^{TT} = \frac{1}{2}\left(I_{xx} - I_{yy}\right) \;\; ; \;\; I_{xy}^{TT} = I_{xy} \qquad \textbf{(4.5.20)}$$

by Fourier analysis. This means that for a gravitational wave travelling in the *z*-direction, the solution to Einstein's linearized equations in the TT-gauge is given by

$$h_{zi}^{TT} = 0 \;\; ; \;\; h_{xx}^{TT} = -h_{yy}^{TT} = -\frac{G}{c^4 r}\left(\ddot{I}_{xx} - \ddot{I}_{yy}\right) \;\; ; \;\; h_{xy}^{TT} = h_{yx}^{TT} = -\frac{2G}{c^4 r}\ddot{I}_{xy} \qquad \textbf{(4.5.21)}$$

where dots denote time-derivatives.

We now introduce *the trace free quadrupole moment tensor* defined by

$$D_{ik} = I_{ik} - \frac{\delta_{ik}}{3} I^j{}_j \qquad \textbf{(4.5.22)}$$

where $I^j{}_j = I_{xx} + I_{yy} + I_{zz}$. At the moment this definition has no significance, but as it will turn out later, it will be of great importance when one tries to find a simple expression for the energy radiated by gravitational radiation (see section 5.1). The solution (4.5.21) for a wave travelling in *z*-direction can now with the help of the trace free quadrupole moment tensor be expressed as

$$h_{zi}^{TT} = 0 \;\; ; \;\; h_{xx}^{TT} = -h_{yy}^{TT} = -\frac{G}{c^4 r}\left(\ddot{D}_{xx} - \ddot{D}_{yy}\right) \;\; ; \;\; h_{xy}^{TT} = h_{xy} = -\frac{2G}{c^4 r}\ddot{D}_{xy} \qquad \textbf{(4.5.23)}$$

Similarly the solutions of Einstein's linearized equations for gravitational waves travelling in the *x*- and the *y*-direction can be found (just by permutation of the *x*,*y* and *z* indexes)

$$h_{xi}^{TT} = 0 \;\; ; \;\; h_{yy}^{TT} = -h_{zz}^{TT} = -\frac{G}{c^4 r}\left(\ddot{D}_{yy} - \ddot{D}_{zz}\right) \;\; ; \;\; h_{yz}^{TT} = h_{yz} = -\frac{2G}{c^4 r}\ddot{D}_{yz} \qquad \textbf{(4.5.24)}$$

$$h_{yi}^{TT} = 0 \;\; ; \;\; h_{zz}^{TT} = -h_{xx}^{TT} = -\frac{G}{c^4 r}\left(\ddot{D}_{zz} - \ddot{D}_{xx}\right) \;\; ; \;\; h_{zx}^{TT} = h_{zx} = -\frac{2G}{c^4 r}\ddot{D}_{zx} \qquad \textbf{(4.5.25)}$$

4.6 The neutron star binary, an example of a gravitational wave source

In this section we will study the neutron star binary as an example of a source to gravitational radiation. Even though there are numerous sources of gravitational waves beside the neutron star binaries, most of these other sources can not be studied with satisfactory accuracy within the linearized theory, since they give rise to very strong gravitational fields.

In order to deal with these sources one must wrestle with the original Einstein equations, which not even the most up to date super-computer is able to do easily. This does not mean that we will avoid studying these strong sources. In chapter 6 we will even study them with the help of the linearized theory and Newtonian theory, hoping to find some qualitative information about these sources.

The planetary orbit in the Newtonian approximation

It is now time for us to continue and calculate the orbits of the neutron stars. Since we are assuming that the gravitational field produced by the binary is weak, and that the stars move much slower than the speed of light, we can use Newtonian theory to calculate the orbits with negligible error.

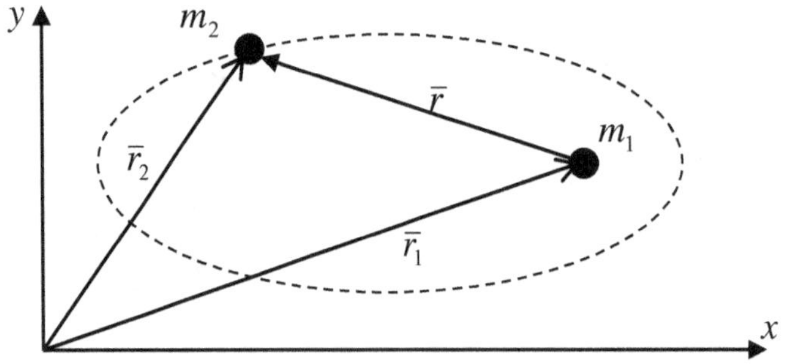

(figure 9)

Let the masses of the stars be m_1 and m_2, then in the centre of mass system the positions of the stars are given by

$$\bar{r}_1^{cm} = -\frac{\mu}{m_1}\bar{r} \quad ; \quad \bar{r}_2^{cm} = \frac{\mu}{m_2}\bar{r} \qquad \textbf{(4.6.1)}$$

where $\mu = \dfrac{m_1 m_2}{m_1 + m_2}$ is the reduced mass of the system, and where $\bar{r} = \bar{r}_2 - \bar{r}_1$ is the relative distance between the stars. The potential energy V and the kinetic energy T of the system are

$$V = -\frac{k}{r} \quad ; \quad T = \frac{\mu \dot{\bar{r}}^2}{2} \qquad \textbf{(4.6.2)}$$

where $k = Gm_1m_2$. In polar coordinates the total energy E and the Lagrangian L are expressed by

$$E = T + V = \frac{\mu}{2}\left[\dot{r}^2 + (r\dot{\theta})^2\right] - \frac{k}{r} \qquad \textbf{(4.6.3)}$$

$$L = T - V = \frac{\mu}{2}\left[\dot{r}^2 + (r\dot{\theta})^2\right] + \frac{k}{r} \qquad \textbf{(4.6.4)}$$

Now we continue by making use of the Lagrange equations

$$\frac{d}{dt}\left(\frac{\partial L}{\partial \dot{q}_i}\right) - \frac{\partial L}{\partial q_i} = 0 \qquad \textbf{(4.6.5)}$$

This give rise to two equations, one for the radius r and one for the angle θ

a) The r-equation

$$\frac{d}{dt}\left(\frac{\partial L}{\partial \dot{r}}\right)-\frac{\partial L}{\partial r}=0 \quad \Leftrightarrow$$

$$\mu\ddot{r}-2\mu r\dot{\theta}^2+\frac{k}{r^2}=0 \tag{4.6.6}$$

b) The θ-equation

$$\frac{d}{dt}\left(\frac{\partial L}{\partial \dot{\theta}}\right)-\frac{\partial L}{\partial \theta}=0 \quad \Leftrightarrow$$

$$\frac{d}{dt}\left(\mu r^2\dot{\theta}\right)=0 \tag{4.6.7}$$

From the θ-equation (4.6.7) we see that the angular momentum l of the system is constant, i.e.

$$l=\mu r^2\dot{\theta}=const. \tag{4.6.8}$$

We can now with the help of (4.6.8) substitute $\dot{\theta}$ with $\frac{l}{\mu r^2}$ in the expression for the total energy, giving the result

$$E=\frac{\mu}{2}\left[\dot{r}^2+\frac{l^2}{\mu^2 r^2}\right]-\frac{k}{r} \tag{4.6.9}$$

The above expression implies that

$$\left(\frac{dr}{dt}\right)^2=\frac{2E}{\mu}+\frac{2k}{\mu r}-\frac{l^2}{\mu^2 r^2} \tag{4.6.10}$$

If one then makes the substitution of variables defined by

$$u=\frac{1}{r} \quad \Rightarrow \quad \frac{dr}{dt}=\frac{dr}{du}\frac{du}{d\theta}\dot{\theta}=-r^2\dot{\theta}\frac{du}{d\theta}=-\frac{l}{\mu}\frac{du}{d\theta} \tag{4.6.11}$$

in the expression (4.6.10) the result becomes

$$\frac{l^2}{\mu^2}\left(\frac{du}{d\theta}\right)^2=\frac{2E}{\mu}+\frac{2k}{\mu}u-\frac{l^2}{\mu^2}u^2 \tag{4.6.12}$$

This implies that

$$\frac{du}{d\theta}=\sqrt{\zeta^2-\left(u-\frac{\mu k}{l^2}\right)^2} \tag{4.6.13}$$

where $\zeta^2=\frac{2E\mu}{l^2}+\left(\frac{\mu k}{l^2}\right)^2$.

The solution to equation (4.6.13) is given by

$$\theta-\theta_0=\int\frac{du}{\zeta\sqrt{1-\frac{1}{\zeta^2}\left(u-\frac{\mu k}{l^2}\right)^2}}=-\cos^{-1}\left[\frac{1}{\zeta}\left(u-\frac{\mu k}{l^2}\right)\right]\Rightarrow$$

$$u=\frac{\mu k}{l^2}+\zeta\cos(\theta-\theta_0) \qquad \textbf{(4.6.14)}$$

where θ_0 is some constant angel. But since $u=r^{-1}$, equation (4.6.14) gives us the planetary orbits in polar coordinates

$$r=\frac{r_0}{1+e\cos(\theta-\theta_0)} \qquad \textbf{(4.6.15)}$$

where $r_0=\frac{l}{k\mu}$, and the eccentricity of the orbits $e=\frac{l^2\zeta}{\mu k}=\sqrt{\frac{2El^2}{\mu k}+1}$. We will assume that the orbits of the binary system are elliptic, i.e., $e<1$.

The quadrupole moment tensor of the binary system

For simplicity, we will here let the neutron stars be modelled as points. This means that the mass distribution of the system is described by

$$\rho(\bar{r})=m_1\delta(\bar{r}-\bar{r}_1^{cm})+m_2\delta(\bar{r}-\bar{r}_2^{cm}) \qquad \textbf{(4.6.16)}$$

where the δ-function is the three-dimensional Dirac-function. This implies that the quadrupole moment tensor is given by

$$I^{ik}=\int_V\left[m_1\delta(\bar{r}-\bar{r}_1^{cm})+m_2\delta(\bar{r}-\bar{r}_2^{cm})\right]x^i x^k\,dV \qquad \textbf{(4.6.17)}$$

The expression of the quadrupole moment tensor then gives us the components of the tensor

$$\begin{aligned}I_{xx}&=m_1\left(x_1^{cm}\right)^2+m_2\left(x_2^{cm}\right)^2\\ I_{yy}&=m_1\left(y_1^{cm}\right)^2+m_2\left(y_2^{cm}\right)^2\\ I_{xy}&=m_1x_1^{cm}y_1^{cm}+m_2x_2^{cm}y_2^{cm}\end{aligned} \qquad \textbf{(4.6.18)}$$

all other components are zero.

From (4.6.1) we get the position-vectors of the stars in the centre of mass system

$$\bar{r}_1^{cm} = \left(x_1^{cm}, y_1^{cm}\right) = -\frac{\mu}{m_1} r(\cos(\theta), \sin(\theta))$$
$$\bar{r}_2^{cm} = \left(x_2^{cm}, y_2^{cm}\right) = \frac{\mu}{m_2} r(\cos(\theta), \sin(\theta)) \qquad \textbf{(4.6.19)}$$

Finally (4.6.18) and (4.6.19) give the components of the quadrupole moment tensor in polar coordinates

$$I_{xx} = \mu r^2 \cos^2(\theta)$$
$$I_{yy} = \mu r^2 \sin^2(\theta) \qquad \textbf{(4.6.20)}$$
$$I_{xy} = \frac{\mu r^2}{2} \sin(2\theta)$$

where the radius r is given by (4.6.15). With the help of the definition (4.5.22) and (4.6.20), it is then easy to calculate the components of the trace free quadrupole moment tensor

$$D_{xx} = \mu r^2 \cos^2(\theta) - \frac{1}{3}\mu r^2$$
$$D_{yy} = \mu r^2 \sin^2(\theta) - \frac{1}{3}\mu r^2$$
$$D_{zz} = -\frac{1}{3}\mu r^2 \qquad \textbf{(4.6.21)}$$
$$D_{xy} = \frac{\mu r^2}{2} \sin(2\theta)$$

The second time-derivative of the trace free quadrupole moment tensor

We have now come to the tricky part of calculating the solution to Einstein's linearized equations, for the case of a binary star system, namely calculating the second time-derivatives of the tensor D_{ik}. In order to simplify these calculations it is convenient to derive the following relation for the time derivative of the radius of the orbit

$$\frac{dr}{dt} = \frac{d}{dt}\left[\frac{r_0}{1 + e\cos(\theta)}\right] = \frac{r_0 e \sin(\theta)}{(1 + e\cos(\theta))^2}\dot{\theta} = \frac{r_0^2}{(1 + e\cos(\theta))^2}\frac{e\sin(\theta)}{r_0}\dot{\theta} = r^2\dot{\theta}\frac{e\sin(\theta)}{r_0} =$$

$$= \mu r^2 \dot{\theta}\frac{e\sin(\theta)}{\mu r_0} \quad \Rightarrow$$

$$\frac{dr}{dt} = \frac{le}{\mu r_0}\sin(\theta) \qquad \textbf{(4.6.22)}$$

a) Derivation of $\frac{d^2}{dt^2}\left(\mu r^2\right)$:

we begin with calculating

$$\frac{d}{dt}\left(\mu r^2\right) = 2\mu r\frac{dr}{dt} = 2\mu r\frac{le}{\mu r_0}\sin(\theta) = \frac{2le\sin(\theta)}{1+e\cos(\theta)} \qquad \textbf{(4.6.23)}$$

then we can continue and use (4.6.23) in the following way

$$\frac{d^2}{dt^2}\left(\mu r^2\right) = \frac{d}{dt}\left[\frac{2le\sin(\theta)}{1+e\cos(\theta)}\right] = 2le\dot{\theta}\frac{\cos(\theta)+e}{(1+e\cos(\theta))^2} = \frac{2le\dot{\theta}r^2}{r_0^{\,2}}(\cos(\theta)+e) \quad \Rightarrow$$

$$\frac{d^2}{dt^2}\left(\mu r^2\right) = \frac{2l^2e}{\mu r_0^{\,2}}(\cos(\theta)+e) \qquad \textbf{(4.6.24)}$$

b) Derivation of $\frac{d^2}{dt^2}\left(\mu r^2\cos^2(\theta)\right)$:

we begin with calculating

$$\frac{d}{dt}\left(\mu r^2\cos^2(\theta)\right) = 2\mu r\frac{dr}{dt}\cos^2(\theta) - 2\mu r^2\dot{\theta}\cos(\theta)\sin(\theta) =$$

$$= 2\mu r\frac{le}{\mu r_0}\cos^2(\theta)\sin(\theta) - 2l\cos(\theta)\sin(\theta) = \frac{2le\cos^2(\theta)\sin(\theta)}{1+e\cos(\theta)} - 2l\cos(\theta)\sin(\theta) \quad \Rightarrow$$

$$\frac{d}{dt}\left(\mu r^2\cos^2(\theta)\right) = -\frac{l\sin(2\theta)}{1+e\cos(\theta)} \qquad \textbf{(4.6.25)}$$

then we continue and use (4.6.23) in the following way

$$\frac{d^2}{dt^2}\left(\mu r^2\cos^2(\theta)\right) = -\frac{d}{dt}\left(\frac{l\sin(2\theta)}{1+e\cos(\theta)}\right) = -l\dot{\theta}\frac{2\cos(2\theta)(1+e\cos(\theta))+e\sin(\theta)\sin(2\theta)}{(1+e\cos(\theta))^2} =$$

$$= -\frac{l^2}{\mu r_0^{\,2}}\left[2\cos(2\theta)(1+e\cos(\theta))+e\sin(\theta)\sin(2\theta)\right] \quad \Rightarrow$$

$$\frac{d^2}{dt^2}\left(\mu r^2\cos^2(\theta)\right) = -\frac{2l^2}{\mu r_0^{\,2}}\left(\cos(2\theta)+e\cos^3(\theta)\right) \qquad \textbf{(4.6.26)}$$

c) Derivation of $\frac{d^2}{dt^2}\left(\mu r^2 \sin(\theta)\right)$

By using methods equivalent to those used in a) and b) it can with some work be shown that

$$\frac{d^2}{dt^2}\left(\mu r^2 \sin(\theta)\right) = -\frac{2l^2}{\mu r_0^2}\left(2e\sin(\theta) + e\sin(2\theta)\cos(\theta) + 2\sin(2\theta)\right) \qquad \textbf{(4.6.27)}$$

Since the second time-derivatives of the trace free quadrupole moment tensor are given by

$$\begin{aligned}
\ddot{D}_{xx} &= \frac{d^2}{dt^2}\left[\mu r^2\cos^2(\theta) - \frac{1}{3}\mu r^2\right] \\
\ddot{D}_{yy} &= \frac{d^2}{dt^2}\left[\mu r^2\sin^2(\theta) - \frac{1}{3}\mu r^2\right] = \frac{d^2}{dt^2}\left[\frac{2}{3}\mu r^2 - \mu r^2\cos^2(\theta)\right] \\
\ddot{D}_{zz} &= -\frac{d^2}{dt^2}\left[\frac{1}{3}\mu r^2\right] \\
\ddot{D}_{xy} &= \frac{d^2}{dt^2}\left[\frac{\mu r^2}{2}\sin(2\theta)\right]
\end{aligned} \qquad \textbf{(4.6.28)}$$

we can now with the help of (4.6.24), (4.6.26) and (4.6.27) get the results of the derivations stated in (4.6.28)

$$\begin{aligned}
\ddot{D}_{xx} &= -\frac{2l^2}{\mu r_0^2}\left[\cos(2\theta) + e\cos^3(\theta) + \frac{1}{3}\left(e\cos(\theta) + e^2\right)\right] \\
\ddot{D}_{yy} &= \frac{2l^2}{\mu r_0^2}\left[\cos(2\theta) + e\cos^3(\theta) + \frac{2}{3}\left(e\cos(\theta) + e^2\right)\right] \\
\ddot{D}_{zz} &= -\frac{2l^2}{3\mu r_0^2}\left(e\cos(\theta) + e^2\right) \\
\ddot{D}_{xy} &= -\frac{l^2}{\mu r_0^2}\left[2e\sin(\theta) + e\sin(2\theta)\cos(\theta) + 2\sin(2\theta)\right]
\end{aligned} \qquad \textbf{(4.6.29)}$$

The solution to the linearized Einstein equations for the neutron star binary

The solution to the linearized Einstein equations for our binary system is now easily calculated with the help of (4.5.23-25) and (4.6.29)

i) Gravitational wave travelling in the z-direction

$$\begin{aligned}
h_{zi}^{TT} &= 0 \\
h_{xx}^{TT} &= -h_{yy}^{TT} = \frac{2G}{c^4 r}\frac{l^2}{\mu r_0^2}\left[2\cos(2\theta) + 2e\cos^3(\theta) + e\cos(\theta) + e^2\right] \\
h_{xy}^{TT} &= \frac{2G}{c^4 r}\frac{l^2}{\mu r_0^2}\left[2e\sin(\theta) + e\sin(2\theta)\cos(\theta) + 2\sin(2\theta)\right]
\end{aligned} \qquad \textbf{(4.6.30)}$$

For circular orbits (*e=0*), these waves are circularly polarized.

ii) Gravitational wave travelling in the x-direction

$$h_{xi}^{TT} = 0$$

$$h_{yy}^{TT} = -h_{zz}^{TT} = -\frac{2G}{c^4 r}\frac{l^2}{\mu r_0^2}\left[\cos(2\theta) + e\cos^3(\theta) + e\cos(\theta) + e^2\right] \quad \textbf{(4.6.31)}$$

$$h_{zx}^{TT} = 0$$

These waves are linearly polarized(observe that the letter *r* appearing in the solutions above denotes the distance from the binary system, and not the relative distance between the two stars).

In this chapter we have seen how gravitational waves effect free particles. The conclusions of these studies will be of uttermost importance when we, in chapter 6, try to explain how the advanced gravitational detectors work. In particular the equations (4.3.13-4.3.14) will be of great help. We have also derived the metric perturbations of gravitational waves emitted by neutron star binaries. The results of these derivations represented by the equations (4.6.30-4.6.31) will later help us to estimate which gravitational wave sources that can be expected to be detected by the previously mentioned detectors.

5 The indirect proof of the existence of gravitational waves

In this chapter we will discuss how, indirectly, one can make the existence of gravitational waves plausible. The method for doing so is based on the prediction that gravitational waves emitted from a source carries away energy from the source. In the case of binary stars, this would mean that the binary system loses energy, making the stars spin faster and closer together. So in measuring the change of the orbits of binary stars, we have the possibility to put the theory of gravitational radiation to the test.

5.1 Energy radiated with gravitational radiation

Since parallel transport of vectors in curved space is dependent on the path, along which we transport our vectors, and since the energy of a particle in a sense is the zero-component of the particle's four-momentum, it is almost impossible to find an invariant definition of energy within the framework of general relativity. However, when dealing with weak gravitational fields these problems can be overcome, making it possible to theoretically derive the average energy radiated per second P, due to the emission of gravitational radiation

$$P = \frac{G}{5c^5}\left\langle \dddot{D}_{ik}\dddot{D}^{ik}\right\rangle \quad \textbf{(5.1.1)}$$

where the brackets denote the time average over one period. It is important to stress the fact that (5.1.1) is an approximation within the linearized theory, only correct to the first order. In this section the deduction of this formula will be discussed.

In order to deduce the formula (5.1.1) describing the energy loss of a gravitational wave source, we must first take a good look at the energy momentum tensor, and interpret its components.

As mentioned before the energy momentum tensor describes the energy and momentum properties of a fluid element, and if we denote the components of this tensor by $T^{\mu\nu}$, these components are to be interpreted in the following way

- T^{00} is the energy density of the fluid element.
- T^{i0} is the p^i-momentum density of the fluid element multiplied with c.
- T^{0k} is the energy flux of the fluid element in the k – direction, divided by c.
- T^{ik} is the p^i-momentum flux of the fluid element in the k – direction.

(Observe that the indexes i and k only run from 1 to 3, i.e. they are spatial indexes.)

We now continue to assume that not only a fluid element but also a gravitational wave has a corresponding energy momentum tensor. If this assumption is correct, we can derive the T^{0k} - component corresponding to a gravitational wave, and by doing that, calculate the energy flux F^k of the gravitational wave, i.e.

$$F^k = cT^{0k} \qquad \textbf{(5.1.2)}$$

Since gravitational waves satisfy the vacuum equations of general relativity

$$G^{\mu\nu} = 0 \qquad \textbf{(5.1.3)}$$

the energy momentum tensor of the gravitational wave must then as a consequence of Einstein's equations be zero, i.e.

$$T^{\mu\nu} = 0 \qquad \textbf{(5.1.4)}$$

Having the relation (5.1.2) in mind it is then clear that the energy flux of the gravitational wave must be zero! However if we study the example of a harmonic oscillator interacting with a gravitational wave (see section 6.1), we observe that the gravitational wave will excite the energy state of the oscillator. So by assuming that the energy flux of the gravitational wave is zero, the conservation of energy is violated (A complete derivation of the energy flux of a gravitational wave, based on the above example has been made by B. F Schutz [13]).

But if we go back to the very first discussion in this section, we see that the conservation of energy in a curved space-time is not something that can be taken for granted. So is it still true that the energy flux of the gravitational wave is zero? The answer to this question despite the implications of the vacuum equations (5.1.3), is no. The motive for this answer derives from the assumption that the disturbances in space-time corresponding to the gravitational wave are small. This means that space-time in the vicinity of the gravitational wave is nearly flat, and that the conservation of energy still can be considered to be true.

When we now have made it plausible that the energy flux of gravitational radiation is different from zero, the next step is to "go around" the implications of the vacuum equations and calculate the energy flux. This is done by the somewhat controversial assumption that the vacuum equations (5.1.3) only are correct to the first order. We can then with the help of Einstein's equations and the non-zero second order components of the Einstein tensor, calculate the energy momentum tensor of the gravitational wave

$$T_{\mu\nu} = \frac{c^4}{8\pi G} G_{\mu\nu} \tag{5.1.5}$$

To be able to calculate the second order components of the Einstein tensor, we will consider the metric of a C_+-polarized gravitational wave travelling in the z-direction, and that is expressed in the TT-gauge

$$g_{\mu\nu} = \eta_{\mu\nu} + h_{\mu\nu}^{TT} = \begin{pmatrix} 1 & 0 & 0 & 0 \\ 0 & -1 + C_+ \cos[\omega(t - z/c)] & 0 & 0 \\ 0 & 0 & -1 - C_+ \cos[\omega(t - z/c)] & 0 \\ 0 & 0 & 0 & -1 \end{pmatrix} \tag{5.1.6}$$

Together with the above metric and the relations (5.1.2) and (5.1.5) the energy flux in the z-direction can be derived

$$F_z = \frac{c^5}{8\pi G} G_{0z} \tag{5.1.7}$$

The only thing we have to do to obtain this energy flux, is to calculate the component G_{0z} to the second order. This calculation is simplified considerably by (1.13.8) which together with (5.1.6) imply

$$G_{0z} = R_{0z} - \frac{g_{0z}}{2} R = R_{0z} \tag{5.1.8}$$

So instead of using (5.1.7) for the calculation of the energy flux, we can use the much simpler relation

$$F_z = \frac{c^5}{8\pi G} R_{0z} \tag{5.1.9}$$

We are now ready to take on the derivation of the energy flux. We start this derivation by expressing the component R_{0z} of the Ricci tensor with the help of (1.11.9) and (1.13.3)

$$R_{0z} = R^{\alpha}{}_{0\alpha z} = \Gamma^{\alpha}_{0z,\alpha} - \Gamma^{\alpha}_{0\alpha,z} + \Gamma^{\sigma}_{0z}\Gamma^{\alpha}_{\sigma\alpha} - \Gamma^{\sigma}_{0\alpha}\Gamma^{\alpha}_{\sigma z} \tag{5.1.10}$$

We then continue and use the relations (1.6.7) and (5.1.6) to make the following derivations

$$\Gamma^{\alpha}_{0z} = \frac{g^{\alpha\sigma}}{2}\left[h^{TT}_{0\sigma,z} + h^{TT}_{\sigma z,0} - h^{TT}_{0z,\sigma}\right] = 0$$

$$\Gamma^{\alpha}_{0\alpha} = \frac{g^{\alpha\sigma}}{2}\left[h^{TT}_{0\sigma,\alpha} + h^{TT}_{\sigma\alpha,0} - h^{TT}_{0\alpha,\sigma}\right] = \frac{g^{\alpha\sigma}}{2} h^{TT}_{\sigma\alpha,0} = \frac{h^{TT}_{xx,0}}{2}\left(g^{xx} - g^{yy}\right)$$

$$\Gamma^{\sigma}_{0\alpha} = \frac{g^{\sigma\lambda}}{2}\left[h^{TT}_{0\lambda,\alpha} + h^{TT}_{\lambda\alpha,0} - h^{TT}_{0\alpha,\lambda}\right] = \frac{g^{\sigma\lambda}}{2} h^{TT}_{\lambda\alpha,0} \tag{5.1.11}$$

$$\Gamma^{\alpha}_{\sigma z} = \frac{g^{\alpha\rho}}{2}\left[h^{TT}_{\sigma\rho,z} + h^{TT}_{\rho z,\sigma} - h^{TT}_{\sigma z,\rho}\right] = \frac{g^{\alpha\rho}}{2} h^{TT}_{\sigma\rho,z}$$

$$\Gamma^{\sigma}_{0\alpha}\Gamma^{\alpha}_{\sigma z} = \frac{\left(g^{\sigma\sigma}\right)^2}{4} h^{TT}_{\sigma\sigma,0} h^{TT}_{\sigma\sigma,z} = \frac{h^{TT}_{xx,0} h^{TT}_{xx,z}}{4}\left[\left(g^{xx}\right)^2 + \left(g^{yy}\right)^2\right]$$

Observe that in the above derivations the identity $h^{TT}_{xx} = -h^{TT}_{yy}$ have been used.

The next step is to use the identity $g^{\mu\sigma} g_{\sigma\nu} = \delta^{\mu}_{\upsilon}$ together with (5.1.6) in order to derive the components of the tensor $g^{\mu\nu}$, which reveals the following two identities

$$g^{xx} = \frac{1}{h^{TT}_{xx} - 1} = -\left(1 + h^{TT}_{xx}\right) + O\left(\left|h^{TT}_{xx}\right|^2\right) \quad ; \quad g^{yy} = \frac{-1}{h^{TT}_{xx} + 1} = h^{TT}_{xx} - 1 + O\left(\left|h^{TT}_{xx}\right|^2\right) \tag{5.1.12}$$

With the help of (5.1.6), (5.1.10), (5.1.11) and (5.1.12) we can now in the second order approximation express the R_{0z}-component of the Ricci tensor

$$R_{0z} = h^{TT}_{xx,0z} h^{TT}_{xx} + \frac{1}{2} h^{TT}_{xx,0} h^{TT}_{xx,z} = \frac{\omega^2 C_+^{\,2}}{c^2}\left(\cos^2[\omega(t - z/c)] - \frac{1}{2}\sin^2[\omega(t - z/c)]\right) \tag{5.1.13}$$

Finally the expression (5.1.13) together with (5.1.9) imply that the energy flux in the z-direction of the gravitational wave is

$$F_z = \frac{c^3}{8\pi G}\omega^2 C_+^{\,2}\left(\cos^2[\omega(t - z/c)] - \frac{1}{2}\sin^2[\omega(t - z/c)]\right) \tag{5.1.14}$$

It is now time to continue towards our final goal of deriving the formula (5.1.1). In order to derive this formula we must first try to relate the energy flux more directly to the metric perturbation of the gravitational wave. The first step in doing so is by calculating the time average of the energy flux over one period

$$\langle F_z \rangle \equiv \frac{\omega}{2\pi}\int_0^{2\pi/\omega} F_z\, dt = \frac{c^3}{32\pi G}\omega^2 C_+^{\,2} \tag{5.1.15}$$

The next step is to calculate the following time average

$$\left\langle h^{TT\,\mu\nu} h^{TT}_{\mu\nu} \right\rangle = \left\langle \left(h^{TT}_{xx}\right)^2 + \left(h^{TT}_{yy}\right)^2 \right\rangle = \frac{\omega}{\pi} \int_0^{2\pi/\omega} \left(h^{TT}_{xx}\right)^2 dt = C_+^{\,2} \qquad \textbf{(5.1.16)}$$

By comparing the expressions (5.1.15) and (5.1.16) it is now easy to see that the average energy flux can be expressed as

$$\left\langle F_z \right\rangle = \frac{c^3}{32\pi G} \omega^2 \left\langle h^{TT\,\mu\nu} h^{TT}_{\mu\nu} \right\rangle \qquad \textbf{(5.1.17)}$$

Observe that even though the expression (5.1.17) has been derived under the assumption that the wave is C_+-polarized, it can easily be shown that this expression is valid for any polarization. In the rest of this derivation we will assume an arbitrary polarization of the gravitational wave.

Since the expression (5.1.1)which we are trying to derive here, contains the trace free quadrupole moment tensor, it is only natural to us to try and reformulate the expression (5.1.7) describing the average energy flux, into an expression containing the very same tensor. This is done by using the identities (4.5.23) describing the solution to the linearized Einstein equations, and since we continue with the assumption that the time dependence of the metric perturbations $h^{TT}_{\mu\nu}$ is sinusoidal, the previously mentioned solutions can be expressed as:

$$h^{TT}_{zi} = 0 \quad ; \quad h^{TT}_{xx} = -h^{TT}_{yy} = \frac{G}{c^4 r} \omega^2 \left(D_{xx} - D_{yy}\right) \quad ; \quad h^{TT}_{xy} = h^{TT}_{yx} = \frac{2G}{c^4 r} \omega^2 D_{xy} \qquad \textbf{(5.1.18)}$$

The expressions (5.1.17) and (5.1.18) now make it possible for us to express the average energy flux in the following way

$$\left\langle F_z \right\rangle = \frac{c^3}{32\pi G} \omega^2 \left\langle \left(h^{TT}_{xx}\right)^2 + \left(h^{TT}_{yy}\right)^2 + 2\left(h^{TT}_{xy}\right)^2 \right\rangle = \frac{G}{16\pi c^5} \frac{\omega^6}{r^2} \left\langle \left(D_{xx} - D_{yy}\right)^2 + 4\left(D_{xy}\right)^2 \right\rangle \qquad \textbf{(5.1.19)}$$

In order to calculate the average energy flux in an arbitrary direction from the gravitational wave source, we now want to replace the expression within the bracket of (5.1.19), with an expression that only has the z-index as a free index. This is done with the help of the identity

$$\left(D_{xx} - D_{yy}\right)^2 + 4\left(D_{xy}\right)^2 = 2D^{ik} D_{ik} - 4D_z^{\ i} D_{iz} + \left(D_{zz}\right)^2 \qquad \textbf{(5.1.20)}$$

(see appendix A for a proof of this identity)

If we now substitute the expression within the bracket of (5.1.19) with (5.1.20), the average energy flux in the z-direction can be expressed as

$$\left\langle F_z \right\rangle = \frac{G}{16\pi c^5} \frac{\omega^6}{r^2} \left\langle 2D^{ik} D_{ik} - 4D_z^{\ i} D_{iz} + \left(D_{zz}\right)^2 \right\rangle \qquad \textbf{(5.1.21)}$$

The beauty of the expression (5.1.21) is that the average energy flux from the source in the radial direction $\hat{n}$, can be calculated just by changing the coordinate basis from Cartesian coordinates to spherical coordinates. By doing that, the average energy flux in the radial direction can be expressed as

$$\langle F_n \rangle = \frac{G}{16\pi c^5}\frac{\omega^6}{r^2}\left\langle 2D^{ik}D_{ik} - 4\tilde{D}_n{}^i\tilde{D}_{in} + \left(\tilde{D}_{nn}\right)^2 \right\rangle \qquad \textbf{(5.1.22)}$$

where $$\tilde{D}_n{}^i\tilde{D}_{in} = D_k{}^i D_{il} n^k n^l \quad ; \quad \tilde{D}_{nn} = D_{kl} n^k n^l \qquad \textbf{(5.1.23)}$$

and where

$$n^x = \sin(\theta)\cos(\phi) \;\; ; \;\; n^y = \sin(\theta)\sin(\phi) \; ; \; n^z = \cos(\theta) \; ; \; \theta \in [0,\pi] \; ; \phi \in [0,2\pi]$$

With the help of (5.1.22-23), we can now by integrating the flux $\langle F_n \rangle$ over the sphere of radius r, obtain the average energy radiated per second P from a gravitational wave source, due to the emission of gravitational radiation, i.e.

$$P = \oiint_S \langle F_n \rangle dS = \frac{G}{16\pi c^5}\frac{\omega^6}{r^2}\left\langle 2D_{ik}D^{ik}\oiint_S dS - 4D_k{}^i D_{il}\oiint_S n^k n^l dS + D_{kl}D_{ij}\oiint_S n^k n^l n^i n^j dS \right\rangle \qquad \textbf{(5.1.24)}$$

The integral in (5.1.24) which does not contain any components of the radial unit vector is just $4\pi r^2$, whereas the other two integrals in the same expression pose a little bigger problem to calculate. In appendix B we show that these integrals are

$$\oiint_S n^k n^l dS = \frac{4\pi}{3} r^2 \delta^{kl} \quad ; \quad \oiint_S n^k n^l n^i n^j dS = \frac{4\pi}{15}\left(\delta^{kl}\delta^{ij} + \delta^{ik}\delta^{jl} + \delta^{il}\delta^{jk}\right) \qquad \textbf{(5.1.25)}$$

Now (5.1.24-25) together with the identity $D^k{}_k = 0$, reveal the following

$$P = \frac{G}{16\pi c^5}\omega^6\left\langle 8\pi D_{ik}D^{ik} - \frac{16\pi}{3}D_k{}^i D_{il}\delta^{kl} + \frac{4\pi}{15}D_{kl}D_{ij}\left(\delta^{kl}\delta^{ij} + \delta^{ik}\delta^{jl} + \delta^{il}\delta^{jk}\right)\right\rangle =$$

$$= \frac{G}{4c^5}\omega^6\left\langle 2D_{ik}D^{ik} - \frac{4}{3}D^{il}D_{il} + \frac{1}{15}\left(D^l{}_l D^i{}_i + D^{ij}D_{ij} + D^{ij}D_{ij}\right)\right\rangle =$$

$$= \frac{G}{4c^5}\omega^6\left\langle D_{ik}D^{ik}\left(2 - \frac{4}{3} + \frac{2}{15}\right)\right\rangle \;\; \Rightarrow$$

$$P = \frac{G}{5c^5}\omega^6\left\langle D_{ik}D^{ik}\right\rangle \qquad \textbf{(5.1.26)}$$

Observe that the expression (5.1.26) has been derived under the assumption that the time dependence of the tensor D_{ik} is sinusoidal. However, with the help of this expression and by

Fourier analysis, it can be shown that when the previously mentioned tensor has a more general time dependence, the expression for the average energy radiated per second, becomes

$$P = \frac{G}{5c^5}\left\langle \dddot{D}_{ik}\dddot{D}^{ik}\right\rangle \qquad \square$$

5.2 The energy radiated by a neutron star binary

We will now use (5.1.1) to calculate the energy radiated per second from a binary star system. In order to do so we must first derive the third time derivatives of the trace free quadrupole moment tensor, which is easily done with the help of (4.6.29)

$$\begin{aligned}
\dddot{D}_{xx} &= \frac{2l^2}{\mu r_0^{\,2}}\left[2\sin(2\theta) + 3e\cos^2(\theta)\sin(\theta) + \frac{e}{3}\sin(\theta)\right]\dot{\theta} \\
\dddot{D}_{yy} &= -\frac{2l^2}{\mu r_0^{\,2}}\left[2\sin(2\theta) + 3e\cos^2(\theta)\sin(\theta) + \frac{2e}{3}\sin(\theta)\right]\dot{\theta} \\
\dddot{D}_{zz} &= \frac{2l^2 e}{3\mu r_0^{\,2}}\dot{\theta}\sin(\theta) \\
\dddot{D}_{xy} &= -\frac{2l^2}{\mu r_0^{\,2}}\left[3e\cos^3(\theta) - e\cos(\theta) + 2\cos(2\theta)\right]\dot{\theta}
\end{aligned} \qquad \textbf{(5.2.1)}$$

The radiated energy per second is then calculated by substituting with (5.2.1) in the integral (5.2.2) below

$$P = \frac{G}{5c^5}\left\langle \dddot{D}_{ik}\dddot{D}^{ik}\right\rangle = \frac{G}{5c^5 T}\int_0^T \left[\left(\dddot{D}_{xx}\right)^2 + \left(\dddot{D}_{yy}\right)^2 + \left(\dddot{D}_{zz}\right)^2 + \left(\dddot{D}_{xy}\right)^2\right]dt \quad \Leftrightarrow$$

$$P = \frac{Gl}{5c^5 T\mu r_0^{\,2}}\int_0^{2\pi}\left[\left(\frac{\dddot{D}_{xx}}{\dot{\theta}}\right)^2 + \left(\frac{\dddot{D}_{yy}}{\dot{\theta}}\right)^2 + \left(\frac{\dddot{D}_{zz}}{\dot{\theta}}\right)^2 + \left(\frac{\dddot{D}_{xy}}{\dot{\theta}}\right)^2\right](1 + e\cos(\theta))^2\, d\theta \qquad \textbf{(5.2.2)}$$

where T is the period.

Performing the integration in (5.2.2) we get that the mean radiated energy per second is given by

$$P = \frac{64Gl^5}{5c^5 T\mu^3 r_0^{\,6}}\left[1 + \frac{73}{24}e^2 + \frac{37}{96}e^4\right] \qquad \textbf{(5.2.3)}$$

By using the well known relations for elliptical orbits

$$r_0 = a(1-e^2)$$
$$l = (k\mu r_0)^{1/2} \tag{5.2.4}$$
$$T = 2\pi a^{3/2}\sqrt{\frac{\mu}{k}} \quad (\textit{Keplers third law})$$

(where a is the length of the principal axis of the ellipse)

the expression (5.2.3) can be given the alternative form

$$P = \frac{32G^4}{5c^5}\frac{(m_1+m_2)^3}{a^5}\frac{\mu^2}{\left(1-e^2\right)^{7/2}}\left[1+\frac{73}{24}e^2+\frac{37}{96}e^4\right] \tag{5.2.5}$$

The change of the distance a, due to the loss of energy

Since the binary system loses energy, the two stars must move closer and closer together, decreasing the length of the orbits principal axis a. The rate at which the principal axis decreases its length can be calculated by using the fact that

$$a = -\frac{k}{2E}$$
$$\left\langle \frac{dE}{dt} \right\rangle = -P \tag{5.2.6}$$

This implies that

$$\frac{da}{dt} = \frac{k}{2E^2}\frac{dE}{dt} = \frac{2a^2}{k}\frac{dE}{dt} \quad \Rightarrow$$

$$\left\langle \frac{da}{dt} \right\rangle = \frac{2a^2}{k}\left\langle \frac{dE}{dt} \right\rangle = -\frac{2a^2}{k}P \tag{5.2.7}$$

Observe that when deriving (5.2.7) it has been assumed that the change in a during one period is very small, i.e. $a(t+T) \approx a(t)$, enabling us to make the approximation $\langle a^2 \rangle \approx a^2$.

If we now substitute P in (5.2.7) with (5.2.5), we will get the relation that tells us how the length of the principal axis changes during one period

$$\left\langle \frac{da}{dt} \right\rangle = -\frac{64G^3}{5c^5}\frac{(m_1+m_2)^2}{a^3}\frac{\mu}{\left(1-e^2\right)^{7/2}}\left[1+\frac{73}{24}e^2+\frac{37}{96}e^4\right] \tag{5.2.8}$$

The change in the period due to the loss of energy

Since the stars in the binary system move closer and closer together, the period of the system must decrease. In order to calculate the rate at which the period decreases, we will use Keplers third law, implying

$$\frac{dT}{dt}=\frac{d}{dt}\left(2\pi a^{3/2}\sqrt{\frac{\mu}{k}}\right)=3\pi a^{1/2}\sqrt{\frac{\mu}{k}}\frac{da}{dt} \tag{5.2.9}$$

This relation then implies under the restriction that a decreases very slowly

$$\left\langle\frac{dT}{dt}\right\rangle=3\pi a^{1/2}\sqrt{\frac{\mu}{k}}\left\langle\frac{da}{dt}\right\rangle \tag{5.2.10}$$

Finally we can then use (5.2.8) and (5.2.10) to obtain the rate at which the period changes

$$\left\langle\frac{dT}{dt}\right\rangle=-\frac{192\pi G^{5/2}}{5c^5}\frac{(m_1+m_2)^{3/2}}{a^{5/2}}\frac{\mu}{\left(1-e^2\right)^{7/2}}\left[1+\frac{73}{24}e^2+\frac{37}{96}e^4\right] \tag{5.2.11}$$

By using Kepler's third law (see 5.2.4), we can substitute a in the above expression with the period T, giving us an alternative version of (5.2.11)

$$\left\langle\frac{dT}{dt}\right\rangle=-\frac{192\pi G^{5/3}}{5c^5}\frac{(m_1+m_2)^{2/3}}{(T/2\pi)^{5/3}}\frac{\mu}{\left(1-e^2\right)^{7/2}}\left[1+\frac{73}{24}e^2+\frac{37}{96}e^4\right] \tag{5.2.12}$$

5.3 Hulse and Taylor

In 1974 Hulse and Taylor discovered a binary system, PSR1913+16, consisting of two very small neutron stars. Since one of these stars is a pulsar, it was possible to measure the change in the period with very god accuracy, as the binary system lost energy. Hulse and Taylor did this, and their measurements revealed that

$$m_1=1.40M_{Solar}\quad;\quad m_2=1.43M_{Solar}\quad;e=0.61724\pm0.00002$$

$$T=27906.98160\pm0.00007\,s\quad;\quad\left\langle\frac{dT}{dt}\right\rangle=(-2.5\pm0.3)\times10^{-12} \tag{5.3.1}$$

The theoretical prediction of the change in the period, calculated with the help of the two star masses, the eccentricity, the period (the above values) and the relation (5.2.12), is

$$\left\langle\frac{dT}{dt}\right\rangle_{Theor.}=(-2.38\pm0.02)\times10^{-12} \tag{5.3.2}$$

This means that the theoretical prediction made by general relativity is consistent with the discovery made by Hulse and Taylor. This was the first measurement that indirectly proved the existence of gravitational waves. The two physicists Hulse and Taylor were awarded the Nobel prize in 1993, for their discovery.

6 Attempts of directly proving the existence of gravitational waves

In this final chapter we will discuss some of the detectors used for detecting gravitational waves, and the sources that can be detected by such detectors. The main focus of this discussion will be around the *Laser Interferometer Gravitational-Wave Observatory* (LIGO), whose construction recently has been finished, and is now being tested. We will also to some extent discuss the early resonant detectors.

6.1 Weber and the resonant detector

In the 1960 Joseph Weber together with scientists from the University of Maryland constructed the first gravitational wave detectors. In this section we will make a simplified discussion of how these detectors work in principle.

The resonant detector can be thought of as two equal masses m separated by a massless spring, with spring constant k, damping constant ν *and* unstretched length l_0. Let us assume that the masses are placed along the *x-axis* of our coordinate system (see figure below)

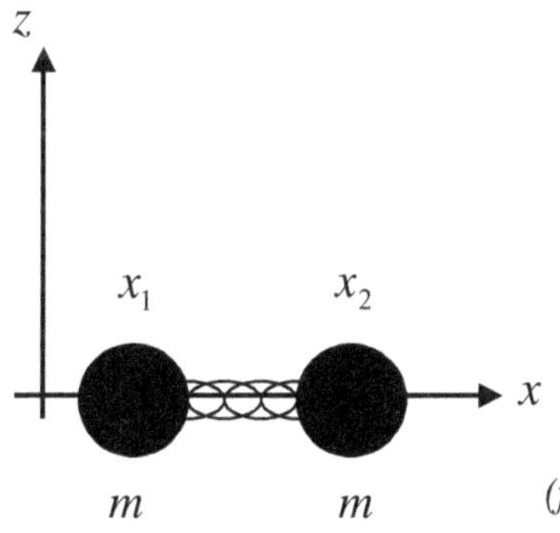

(figure 10)

If u denotes the elongation of the spring, then the positions of the masses in the centre of mass system are given by

$$x_1 = \frac{1}{2}(l_0 - u) \;\; ; \;\; x_2 = \frac{1}{2}(l_0 + u) \qquad \textbf{(6.1.1)}$$

Let the separation vector between the centre of mass to the mass m_2 be denoted by ξ^α, i.e.

$$\xi^\alpha = \frac{1}{2}(0, l_0 + u, 0, 0) \qquad \textbf{(6.1.2)}$$

Let us suppose that a gravitational wave travelling in the z-direction "hits" the detector, then the movement of the right mass in the Newtonian approximation would be described by the equation

$$m\frac{d^2 x_2}{dt^2} = -ku - \nu\frac{du}{dt} + F \quad \Leftrightarrow$$

$$\frac{m}{2}\frac{d^2 u}{dt^2} = -ku - \nu\frac{du}{dt} + F \qquad \textbf{(6.1.3)}$$

where F is the tidal force induced on the right mass, by the gravitational wave, occurring when the wave changes the distance between the mass and the centre of mass. Since the tidal acceleration is given by the equation of geodesic deviation (1.12.10), the tidal force in the Newtonian approximation, can be expressed by

$$F = m\frac{D^2\xi^\alpha}{dt^2} = mR^\alpha{}_{\beta\mu\sigma}U^\beta U^\mu \xi^\sigma \qquad \textbf{(6.1.4)}$$

Since the velocities of the masses are much smaller than the speed of light, and since the elongation is much smaller than the unstretched length ($|u| << l_0$), the expression above can to the first order be approximated as

$$F = \frac{mc^2 l_0}{2} R^x{}_{00x} \qquad \textbf{(6.1.5)}$$

According to the expression (3.1.4) for the Riemann tensor in the linear approximation, we get that

$$R^x{}_{00x} = \frac{\eta^{x\sigma}}{2}\left(h^{TT}_{\sigma x,00} + h^{TT}_{00,\sigma x} - h^{TT}_{\sigma 0,0x} - h^{TT}_{0x,\sigma 0}\right) = -\frac{1}{2c^2}\frac{\partial^2 h^{TT}_{xx}}{\partial t^2} \qquad \textbf{(6.1.6)}$$

This means that the tidal force acting on the mass is given by

$$F = -\frac{ml_0}{4}\frac{\partial^2 h^{TT}_{xx}}{\partial t^2} \qquad \textbf{(6.1.7)}$$

Substituting in equation (6.1.3) with (6.1.7), we get the equation for the forced harmonic oscillator

$$\frac{d^2u}{dt^2} + 2\gamma\frac{du}{dt} + \omega_0{}^2 u = -\frac{l_0}{2}\frac{\partial^2 h^{TT}_{xx}}{\partial t^2} \qquad \textbf{(6.1.8)}$$

where $\gamma = \frac{\nu}{m}$ and $\omega_0{}^2 = \frac{2k}{m}$.

We now assume that the metric perturbation of the gravitational wave is described by

$$h^{TT}_{xx} = h\cos(\Omega t) \qquad \textbf{(6.1.9)}$$

where Ω is the frequency of the wave and h the amplitude of the wave. Then by substituting with (6.1.9) in equation (6.1.8) we get the equation

$$\frac{d^2u}{dt^2} + 2\gamma\frac{du}{dt} + \omega_0{}^2 u = \frac{h\Omega^2 l_0}{2}\cos(\Omega t) \qquad \textbf{(6.1.10)}$$

The homogeneous solution to equation (6.1.10)

Through basic methods used in the theory of ordinary differential equations, the homogeneous solution to (6.1.10) is easily found

$$u_h = e^{-\gamma t}\left[A\sin\left(t\sqrt{\omega_0^{\,2}-\gamma^2}\right)+B\cos\left(t\sqrt{\omega_0^{\,2}-\gamma^2}\right)\right] \tag{6.1.11}$$

where A and B are constants, which can be determined by the initial condition of the detector.

The particular solution to equation (6.1.10)

This solution is found by assuming that the solution has the form

$$u_p = C\cos(\Omega t)+D\sin(\Omega t) \tag{6.1.12}$$

where C and D are constants. Then by substituting with (6.1.12) in equation (6.1.10) one gets the relation

$$\left(\omega_0^{\,2}C+2\gamma\Omega D-\Omega^2 C\right)\cos(\Omega t)+\left(\omega_0^{\,2}D-2\gamma\Omega C-\Omega^2 D\right)\sin(\Omega t)=\frac{h\Omega^2 l_0}{2}\cos(\Omega t) \tag{6.1.13}$$

If the above expression is to be true for all t, the following two equations for the constants C and D must be satisfied

$$\begin{aligned}\left(\omega_0^{\,2}-\Omega^2\right)C+2\gamma\Omega D &= \frac{h\Omega^2 l_0}{2}\\ \left(\omega_0^{\,2}-\Omega^2\right)D-2\gamma\Omega C &= 0\end{aligned} \tag{6.1.14}$$

The solution to this modest system of equations is

$$C=\frac{h\Omega^2 l_0}{2}\frac{\omega_0^{\,2}-\Omega^2}{\left(\omega_0^{\,2}-\Omega^2\right)^2+4(\gamma\Omega)^2} \quad ; \quad D=\frac{h\Omega^2 l_0}{2}\frac{2\gamma\Omega}{\left(\omega_0^{\,2}-\Omega^2\right)^2+4(\gamma\Omega)^2} \tag{6.1.15}$$

This means that the particular solution is given by

$$u_p=\frac{h\Omega^2 l_0}{2}\frac{1}{\left(\omega_0^{\,2}-\Omega^2\right)^2+4(\gamma\Omega)^2}\left[\left(\omega_0^{\,2}-\Omega^2\right)\cos(\Omega t)+2\gamma\Omega\sin(\Omega t)\right]=R\sin(\Omega t+\phi) \tag{6.1.16}$$

where

$$R=\frac{h\Omega^2 l_0}{2\sqrt{\left(\omega_0^{\,2}-\Omega^2\right)^2+4(\gamma\Omega)^2}} \quad ; \quad \tan(\phi)=\frac{\omega_0^{\,2}-\Omega^2}{2\gamma\Omega} \tag{6.1.17}$$

Finally by assuming that $t >> \gamma^{-1}$ we can expect that all transients have died away, which means that we don't have to include the homogeneous solution to the solution of (6.1.10) implying that the final solution is

$$u = u_p = R\sin(\Omega t + \phi) \qquad \textbf{(6.1.18)}$$

The energy of the resonance detector

The energy in the resonance detector due to the excitation by the gravitational wave is given by

$$E = \frac{m}{2}\left[\left(\frac{dx_1}{dt}\right)^2 + \left(\frac{dx_2}{dt}\right)^2\right] + \frac{ku^2}{2} \qquad \textbf{(6.1.19)}$$

By substituting with (6.1.1) in (6.1.19) we get the alternative and more useful relation

$$E = \frac{m}{4}\left(\frac{du}{dt}\right)^2 + \frac{ku^2}{2} \qquad \textbf{(6.1.20)}$$

Substituting with (6.1.18) in the above expression we get our final expression for the energy of the detector induced by the gravitational wave

$$E = \frac{mR^2}{4}\left[\Omega^2 \cos^2(\Omega t + \phi) + \omega_0^{\ 2} \sin^2(\Omega t + \phi)\right] \qquad \textbf{(6.1.21)}$$

Resonance

If the incoming gravitational wave has the same frequency as the resonance frequency of the detector, i.e. $\Omega = \omega_0$, then the amplitude and the energy of the detector will be

$$R_{resonance} = \frac{h\Omega l_0}{4\gamma}$$

$$E_{resonance} = \frac{mh^2\Omega^2 l_0^{\ 2}}{16}\left(\frac{\Omega}{2\gamma}\right)^2 = \frac{mh^2\Omega^2 l_0^{\ 2}}{16} Q^2 \qquad \textbf{(6.1.22)}$$

where $Q = \Omega / 2\gamma$ is the so called quality-factor of the detector.

Example

As an illustrating example let us assume that our detector is one of the first resonance detectors built by J.Weber and his team. These detectors were aluminium bars with the following specifications[14]

$$m = 1.4\times 10^3\,kg \ ; \ l_0 = 1.5m \ ; \ \omega_0 = 10^4\,rad\cdot s^{-1} \ ; \ Q = 10^5 \qquad \textbf{(6.1.23)}$$

Further more let us also assume that a strong gravitational wave with amplitude $h = 10^{-20}$ and the same frequency as the detectors resonant frequency "hits" the detector. Then by using these data and the specifications given in (6.1.23) the excitation energy can be calculated with the help of (6.1.22)

$$E_{resonance} \approx 2 \times 10^{-20} J \quad \textbf{(6.1.24)}$$

Comparing this energy with mean energy of the thermal noise at room temperature (300 *K*), in the detector

$$E_{Termal} = KT \approx 4 \times 10^{-21} J \quad \textbf{(6.1.25)}$$

(where *K* is the Boltzmann's constant).
It becomes clear that the detector does not only need to be shielded from mechanical vibrations coming from the surroundings, but it must also be properly cooled, if it is to detect even the strongest gravitational waves.

Other problems that will appear when one tries to use a resonant detector comes from the prediction that most of the strongest gravitational waves that reach Earth come in burst. These waves are released from violent astrophysical events such as supernovae explosions and the collapse of neutron stars in to black holes, not lasting long enough to excite the detector to its full resonant amplitude. Another limitation is that the detectors are only sensitive to gravitational waves with frequencies near the detectors resonance frequency.

Despite these problems, J.Weber claims that the measurements made with these resonant detectors prove the existence of gravitational waves. However, other scientist who have been using similar detectors are not quite as optimistic regarding the outcome of these experiments[15].

6.2 LIGO

The Laser-Interferometer Gravitational-wave project, involves the building and running of two large-scale interferometers in the United States. One is situated in Livingston, Louisiana, and other one in Hanford, Washington. The objective of this project is to directly detect gravitational waves.

The construction of the interferometers is now complete, and a series of engineering runs have now been launched in order to make the interferometers fit for gravitational wave scanning in 2002 [16].

6.2.1 The interferometers

Conceptually the equipment used by LIGO are two large state of the art Michelson-Interferometers. A simplified sketch of the optical layout of the interferometers is shown below[17].

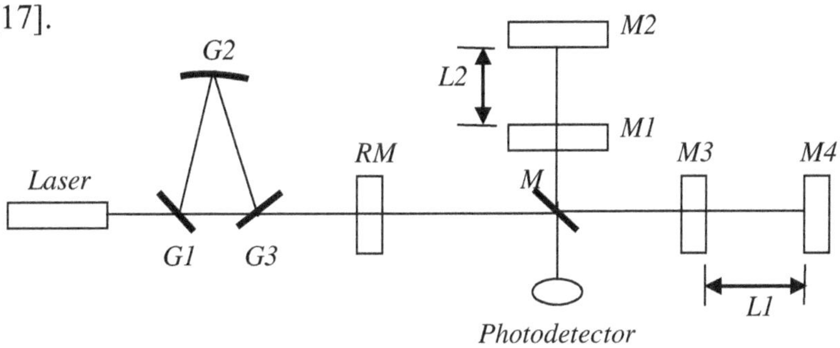

(figure 11)

- *The Laser*
  The heart of the equipment is a Nd:YAG laser ($\lambda \approx 10^3 nm$), together with electro-optic components which stabilize the frequency and the intensity of the laser beam.

- *The mode cleaner*
  In order to get a Gaussian distribution on the different modes of the laser beam as it enters the interferometer, a mode cleaner is used. The mode cleaner is a triangular cavity constructed of three mirrors *G1-G3* (see figure 11), where the mirrors *G1* and *G3* are partially transparent.

- *The recycling mirror*
  To increase the power of the laser light entering the interferometer, a partially reflective mirror *RM* (see figure 11) is used. This mirror recycles the light exiting the interferometer by reflecting it back into the interferometer in phase with the light that is entering the interferometer for the first time.

- *The test-mass-mirrors*
  In each end and beginning of the two interferometer arms *L1* and *L2* (see figure 11) large test-mass-mirrors *M1-M4* (see figure 11) have been placed. These mirrors are suspended by single loop wires. The positions of these mirrors are actively adjusted by electrical coils which induces force on magnets placed on the back of the mirrors.

- *The interferometer arms*
  The two orthogonal optical cavities *L1* and *L2* (se figure 11) are the interferometer arms. When a gravitational wave "hits" the interferometer, the wave will induce a length difference between the mirrors in the two arms, and it is with the help of this length difference one hopes to detect gravitational waves. The interferometer arms in Livingston and in Louisiana are 2km respectively 4km long.

- *The photo-detector*
  After the two split beams have been reflected several times in their respective arm cavities, they will be realigned on the photo-detector, which detects whether the two beams are in or out of phase with each other.

6.2.2 How LIGO operates

The operation of the interferometers includes controlling several degrees of freedom. The parameters representing this freedom are the laser frequency, alignment of the test mass mirrors and the lengths of the different optical cavities of the interferometer.
The control over the parameters mentioned is obtained by a multitude of interconnected servo systems, which keep the interferometer "locked" on a dark spot on the photo diode (i.e. keeping the two laser-beams entering the diode 180 degrees out of phase).

When a gravitational wave "hits" the interferometer a length difference will be induced between the two arms and the wave will be detected as an effort by the servo system, trying to conserve the destructive interference by compensating for the length difference between the two arms.

As an illustrating example of how the interferometers work, we will assume that a C_+-polarized wave travelling in the z-direction "hits" the interferometer. For simplicity we will also assume that the x and y-directions of the wave's TT-coordinate system coincide with the two interferometer arms, and that the source is located right above the interferometer.(In the remainder of this paper we will assume that the source is located and directed in this manner. This orientation and direction of the source will be referred to as the optimal orientation and direction.) Furthermore, let the metric perturbation of the wave be represented by

$$h(t) = h\cos(\omega t - k_0 z) \qquad \textbf{(6.2.2.1)}$$

where $h(t) = h_{xx}^{TT} = -h_{yy}^{TT}$ and $h = C_+$. Let the lengths of the two interferometer arms be denoted by

$$L1 = \xi^1(t) \quad ; \quad L2 = \xi^2(t) \qquad \textbf{(6.2.2.2)}$$

The initial distances $\xi^1(0)$ and $\xi^2(0)$ between the test mass mirrors in the respective interferometer arms, will here be equal to the arm length, i.e.

$$\xi^1(0) = \xi^2(0) = L \qquad \textbf{(6.2.2.3)}$$

Now by making use of the theory developed in section 4.3 and the equations 4.3.13-14, it is easy to express the lengths of the arms when the interferometer is under the influence of a gravitational wave

$$\begin{aligned} \xi^1(t) &= L\left(1 - \frac{h(t)}{2}\right) \\ \xi^2(t) &= L\left(1 + \frac{h(t)}{2}\right) \end{aligned} \qquad \textbf{(6.2.2.4)}$$

Consequently the fractional length difference between the interferometer arms will be

$$\Delta L = \xi^2(t) - \xi^1(t) = Lh(t) \quad \Rightarrow$$

$$\frac{\Delta L}{L} = h(t) \qquad \textbf{(6.2.2.5)}$$

This means that the maximum fractional length difference induced by a gravitational wave, will be equal to the amplitude of the wave.

The typical amplitudes of the gravitational waves expected to be detected by LIGO in its first operative stages are $h \geq 10^{-21}$ [19]. This implies that the interferometers must be able to detect fractional length changes of order 10^{-21}, or equivalently in the case of the detectors placed in Livingston, detect length differences equal to

$$\Delta L = hL \approx 4000 \cdot 10^{-21} m = 4 \cdot 10^{-18} m \qquad \textbf{(6.2.2.6)}$$

which is 1/1000 the diameter of the nucleus of an atom! In the following two sections the problem of detecting such small length differences, and how LIGO deals with these problems will be discussed.

6.2.3 Different noise sources and how they limit the sensitivity of LIGO

The limitations to the interferometers sensitivity arise from the fact that the detector is subject to different noise sources. The main noise sources will be discussed below[20].

- *Seismic noise*
  The source of this noise is the seismic vibrations in the ground beneath the interferometer. These vibrations will make the test mass mirrors move back and forth, and thus impersonate the appearance of a gravitational wave in the interferometer. The strongest disturbances from this noise source will appear for frequencies $f \leq 40Hz$.

- *Thermal noise*
  The source of this noise is the thermal vibrations of the atoms in the test mass mirrors and in the wires that suspend these mirrors. The main contribution to the noise from this source will be in the frequency domain $40Hz < f < 200Hz$.

- *Photon shot noise (Phase noise)*
  The source of this noise is the random arrival times of the photons reaching the photo diode. The strongest disturbances from this noise source will appear for frequencies $f \geq 200Hz$

6.2.4 How LIGO deals with the noise problem

The next fundamental and natural question to ask is how one can expect to measure length differences as small as 1/1000 the diameter of the nucleus of an atom, despite the noise problems mentioned previously. To answer this question, we will, for the rest of this section, follow the main outline of a simplified discussion regarding the noise problem at LIGO, made by Kip S. Thorne[21].

Amplification of the phase shift

The firs step in overcoming the noise problem is to amplify the phase shift between the laser beams in the two interferometer arms. This is done by adjusting the reflectivity of the interferometer mirrors *M1* and *M2* (see figure 11), so that under half a period of a gravitational wave with frequency $\approx 100Hz$, the laser light will be contained in the interferometer arm cavities for an average of 100 round trips. This means that the total phase shift would roughly be

$$\Delta\phi \approx 100 \cdot 2 \cdot \frac{2\pi\Delta L}{\lambda} \approx 10^{-9} \qquad \textbf{(6.2.4.1)}$$

So in this way, the phase shift between the two laser beams entering the photo detector, is amplified approximately 100 times.

Dealing with the photon noise problem

The accuracy which can be obtained when measuring the phase shift between the two realigned laser beams is given by the relation

$$\Delta\phi \approx \frac{1}{\sqrt{N}} \qquad \textbf{(6.2.4.2)}$$

where N is the number of photons entering the interferometer arms during the half period of storage time[22]. To obtain the required accuracy $\Delta\phi = 10^{-9}$, the number of photons entering the arm cavities must thus be

$$N > \frac{1}{\Delta\phi^2} \approx 10^{18} \qquad \textbf{(6.2.4.3)}$$

This means that in order to acquire sufficient accuracy, the laser must provide the arm cavities with more than 10^{18} photons in 0.005 second. This implies that the power P of the laser must be

$$P > \frac{hcN}{0.005\lambda} \approx 40W \qquad \textbf{(6.2.4.4)}$$

(where h is Planck's constant and $\lambda = 1\mu m$ is the wave length of the laser light)
So by keeping the power of the laser beam (the beam travelling towards the mirror M in figure11) around 100W, the undesired photon shot noise can be kept under control. This rather high laser power is obtained by means of the recycling mirror described in section 6.2.1.

Dealing with the thermal noise problem

The length fluctuations due to the thermal vibrations of the atoms in the mirror surfaces, can be estimated by assuming that the atoms move in small circles of radius $\Delta l_{thermal}$ with angular frequency ω. This means that the kinetic energy E_{Kin} of an individual atom due to these circular vibrations is given by

$$E_{kin} = \frac{m(\omega\Delta l_{thermal})^2}{2} \qquad \textbf{(6.2.4.5)}$$

(where m is the atomic rest mass)
Since this energy must be equal to the thermal energy $\frac{kT}{2}$ of this specific vibration mode, we will have

$$\Delta l_{thermal} = \sqrt{\frac{kT}{m\omega^2}} \approx 10^{-12} m \qquad \textbf{(6.2.4.6)}$$

(where k is Boltzmann's constant and T is room temperature (300K), $\omega = 10^{14} s^{-1}$)

This thermally induced length fluctuation is 10^6 times larger than the length change created by the gravitational waves. But this is not really a big problem. The reason for this comes mainly from the fact that the laser beam contained in the interferometer arms has a diameter of $\approx 5cm$, and thus will be reflected of approximately 10^{17} atoms on the surface of the mirror. The second reason comes from the fact that the beam is stored in the arm cavity in 0.005 sec, enabling it to interact with an average of 10^{11} vibrations of each individual atom. Hence it is not very hard to realize that the average value of the thermally induced length fluctuations will be negligible.

The thermal noise due to the thermal vibrations in the suspension wires will pose a much bigger problem. For a full discussion, we refer to the work of Constantin Brif [23]

Dealing with the seismic noise problem

The seismic vibrations at the LIGO locations make the test mass mirrors move, and thus induce fluctuations in the distance between the mirrors. The length fluctuations due to these vibrations are of the order

$$\Delta l_{seismic} = 10^{-8}\,\text{m} \qquad \textbf{(6.2.4.7)}$$

and the frequency range of these vibrations is $f < 100Hz$. In order to protect the interferometer from the seismic disturbances, the wires suspending the test mass mirrors have been attached to the ceiling of the arm cavity through a layer of harmonic oscillators. Each of these oscillators have a resonant frequency $f_0 \approx 1Hz$. When the seismic vibrations try to drive these harmonic oscillators, each oscillator will damp these vibrations by an approximate factor $\left(\frac{f_0}{f}\right)^2$, where f is the frequency of the seismic vibrations. This implies that if the seismic vibrations have frequencies around and above 10Hz, each oscillator will damp the amplitude of the seismic noise by a factor ≥ 100. So by using a stack of say 7 oscillators, the total damping will be of a factor 10^{14}, and the total length fluctuation would thus be

$$\Delta l_{seismic} \approx 10^{-22}\,m \qquad \textbf{(6.2.4.8)}$$

which is more than acceptable for acquiring the desired sensitivities.

After the prescribed noise discriminating measures have been applied, the sensitivity of the LIGO interferometers is described by the following diagram

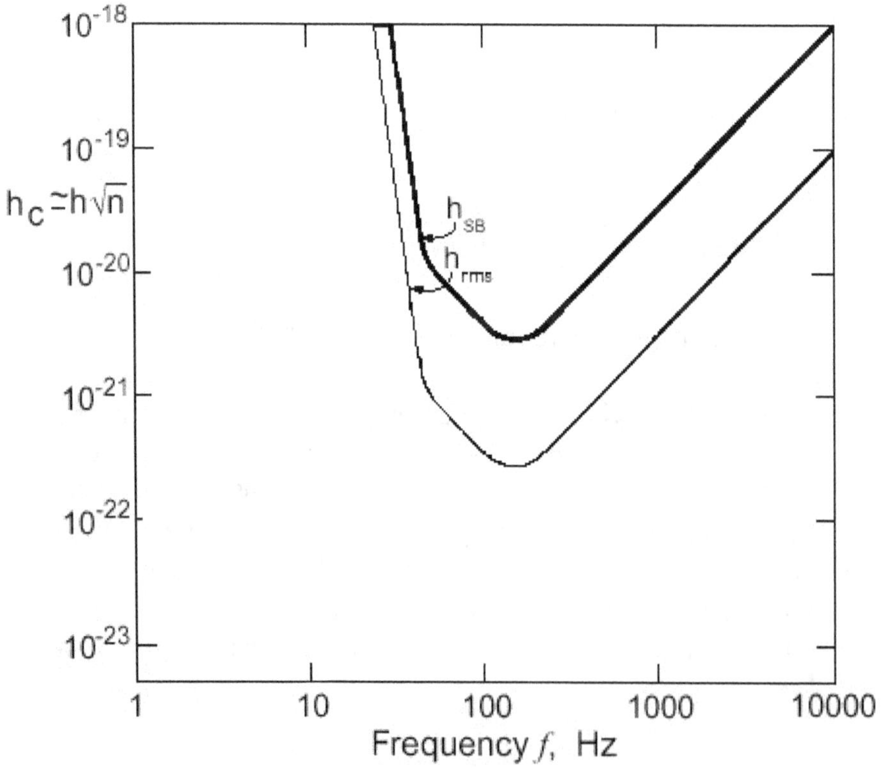

(figure 12)

In figure 12 (previous page) the characteristic signal strength h_c has been plotted as a function of the frequency. Where n is the number of periods a gravitational wave source emits waves of a specific frequency, h_{rms} is the broad band noise and h_{SB} is the sensitivity to burst[24].

6.2.5 The sources expected to be detected by LIGO

In this section we will discuss some of the sources that might be detected by the LIGO interferometers. As an introductory example we will study the neuron binary PSR1913+16, in order to determine if LIGO can detect the gravitational radiation emitted from this binary.

Can the radiation from PSR1913+16 be detected ?

As a first step in answering this question, we assume for the sake of simplicity that the two neutron stars travel in a circular orbit around their common centre of mass, i.e. $e = 0$. For a wave travelling in the z-direction we then get from equation (4.6.30) that the metric perturbation is given by

$$h_{xx}^{TT} = -h_{yy}^{TT} = \frac{4G}{c^4 r}\frac{l^2}{\mu r_0^2}\cos(2\theta)$$
$$h_{xy}^{TT} = \frac{4G}{c^4 r}\frac{l^2}{\mu r_0^2}\sin(2\theta) \qquad \textbf{(6.2.5.1)}$$

Furthermore the circular orbits the following relations are also true

$$l^2 = \mu^2 a^4\left(\frac{2\pi}{T}\right)^2 \;;\; \theta = \frac{2\pi t}{T} \;;\; ;r_0 = a \;;\; T = 2\pi a^{3/2}\sqrt{\frac{\mu}{k}} \qquad \textbf{(6.2.5.2)}$$

making it possible to express the metric perturbation of the waves in the more simple form

$$h_{xx}^{TT} = -h_{yy}^{TT} = \frac{(16\pi)^{2/3}G^{5/3}\mu}{c^4 r}\left(\frac{m_1+m_2}{T}\right)^{2/3}\cos\left(\frac{4\pi t}{T}\right)$$
$$h_{xy}^{TT} = \frac{(16\pi)^{2/3}G^{5/3}\mu}{c^4 r}\left(\frac{m_1+m_2}{T}\right)^{2/3}\sin\left(\frac{4\pi t}{T}\right) \qquad \textbf{(6.2.5.3)}$$

By using the data of PSR1913+16 , i.e. the data given by (5.2.1) together with the distance between the binary and Earth $r = 5kpc$, the amplitude h and the frequency f of the gravitational waves can be estimated

$$f = \frac{2}{T} \approx 70mHz$$
$$h = \frac{(16\pi)^{2/3}G^{5/3}\mu}{c^4 r}\left(\frac{m_1+m_2}{T}\right)^{2/3} \approx 6\cdot 10^{-20} \qquad \textbf{(6.2.5.4)}$$

Comparing these data with the diagram describing the sensitivities for the LIGO interferometers (figure 12), the conclusion must be that the gravitational radiation emitted by PSR1913+16 cannot be detected by LIGO. The reason for this is obviously that the binary emits gravitational waves with a frequency that is too low for LIGO to detect. But what happens when the orbits of the binary decay due to the loss of energy by emission of gravitational radiation? Surely, the stars must spiral inward, increasing the amplitude and the frequency of the emitted gravitational waves, so that at some point the frequency of the waves is high enough for LIGO to detect them. To study this process we begin with the expression (5.1.13) describing the decay of the period

$$\left\langle \frac{dT}{dt} \right\rangle = -\frac{192\pi G^{5/3}}{5c^5} \frac{\mu\,(m_1+m_2)^{2/3}}{(T/2\pi)^{5/3}} \quad \textbf{(6.2.5.5)}$$

(We continue with the assumption that the orbits are circular)

By assuming that the decay is slow, i.e. $\left\langle \frac{dT}{dt} \right\rangle \approx \frac{dT}{dt}$, the equation (6.2.5.5) can be written in the form of a simple ordinary differential equation

$$\frac{dT}{dt} = -\frac{A}{T^{5/3}} \quad \textbf{(6.2.5.6)}$$

where $A = \frac{192\pi(G2\pi)^{5/3}\mu}{5c^5}(m_1+m_2)^{2/3}$. The solution of this ordinary differential equation is given by

$$\int T^{5/3}dT = -\int A dt \quad \Rightarrow \quad \frac{3}{8}T^{8/3} = B - At \quad \Rightarrow$$

$$T = \left(\frac{8}{3}\right)^{3/8}(B-At)^{3/8} \quad \textbf{(6.2.5.7)}$$

(where B is an constant of integration) By using the initial condition

$$T(0) = T_0 = 27906.98\ s \quad \textbf{(6.2.5.8)}$$

The constant B can be determined, revealing the final solution

$$T(t) = T_0\left(1 - \frac{8At}{3T_0^{\;8/3}}\right)^{3/8} \quad \textbf{(6.2.5.9)}$$

By substituting the period T in equation (6.2.5.3) with (6.2.5.9) we get that the metric perturbation of the gravitational waves emitted by PSR1913+16 can be expressed as

$$h_{xx}^{TT} = -h_{yy}^{TT} = \frac{(16\pi)^{2/3} G^{5/3} \mu}{c^4 r} \left[\frac{m_1 + m_2}{T_0 \left(1 - \frac{8At}{3T_0^{8/3}}\right)^{8/3}} \right]^{2/3} \cos\left(\frac{4\pi t}{T_0 \left(1 - \frac{8At}{3T_0^{8/3}}\right)^{8/3}} \right)$$

(6.2.5.10)

$$h_{xy}^{TT} = \frac{(16\pi)^{2/3} G^{5/3} \mu}{c^4 r} \left[\frac{m_1 + m_2}{T_0 \left(1 - \frac{8At}{3T_0^{8/3}}\right)^{8/3}} \right]^{2/3} \sin\left(\frac{4\pi t}{T_0 \left(1 - \frac{8At}{3T_0^{8/3}}\right)^{8/3}} \right)$$

We can now use the above expressions (6.2.5.10) to plot the *xx*- component of the metric perturbation.

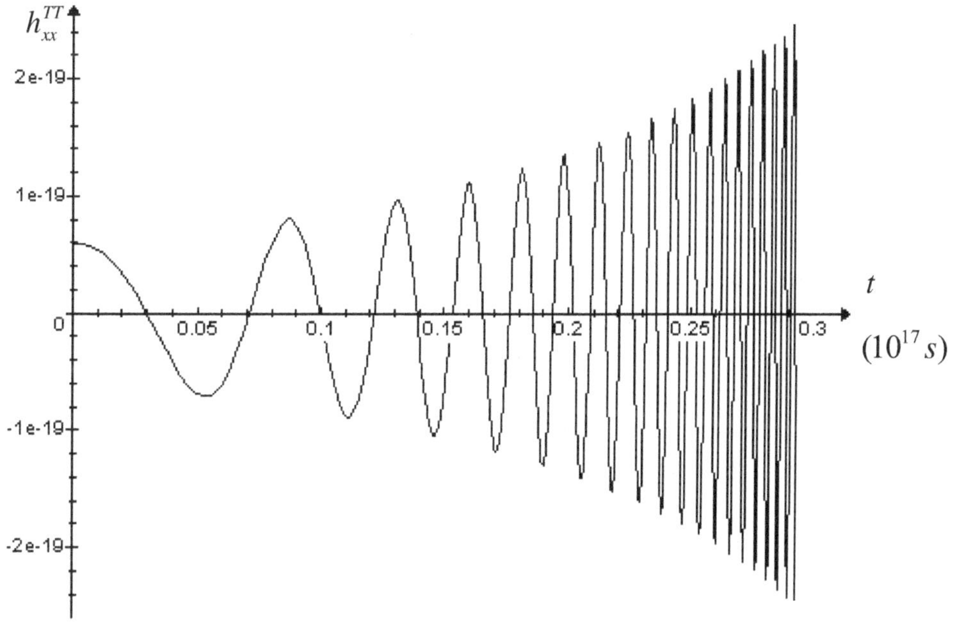

(figure 13)

Observe that in the graph above (figure 13), the periods of the gravitational wave have been made 10^{12} times bigger than the actual ones (i.e. the arguments of the cosine and sin-functions in the expressions stated in (6.2.5.10) have been multiplied with a factor 10^{-12}). So the information that can be extracted from this graph is of strictly conceptual nature.

The graph in figure 13, now confirms the assumption that the frequency and the amplitude of the gravitational waves emitted by the binary will increase. The next question is how long it takes the binary to start emitting gravitational waves that can be detected by LIGO? From the sensitivity curve in figure 12, we can see that LIGO is most sensitive to waves near 100Hz. This means that when the orbit of PSR1913+16 has decayed long enough to emit waves with a frequency of around 100Hz, these waves will be detected by LIGO. To calculate how long this decay will take, we rewrite the expression (6.2.5.9) in the following way

$$t = \frac{3}{8A}\left(T_0^{8/3} - T^{8/3}\right) \qquad \textbf{(6.2.5.11)}$$

(Where t is the time it takes for the period to decay from T_0 to T), and by letting $T = 0.02s$, i.e. a period corresponding to a wave with frequency 100Hz, the decay time can be estimated to

$$t \approx 5.3 \cdot 10^{16} s \approx 1.7 \cdot 10^{9}\ years \qquad \textbf{(6.2.5.12)}$$

In other words we cannot expect LIGO to detect any gravitational waves in the near future, emitted from the binary PSR1913+16. This rather rough estimate can also be compared with better estimates which indicate a decay time of $10^8\ years$[25].

Sources that at present date can be expected to be detected by LIGO

Finally we take a look at some of the gravitational wave sources, which can be expected to be detected by LIGO in a near future. The sources that we will study here are black hole and neutron star binaries that spiral inward along circular orbits and in the end coalesce. In this study we will use very rough estimates that are based on the quadrupole approximation (linearized theory), and the Newtonian approximation. In order to determine whether or not these sources can be detected by LIGO, we will calculate the characteristic signal strengths $h_c = h\sqrt{n}$ for the waves emitted by the previously mentioned binaries, and compare them with the LIGO sensitivities. But before we carry on, we will try to explain why the LIGO sensitivities are being compared with the characteristic signal strengths, instead of just comparing them with the amplitudes of the gravitational waves. The reason for this, is that the noise at a specific frequency, only contributes to the output signal of LIGO with a limited number of periods per second. So the more periods a gravitational wave source emits waves with a certain frequency, the more probable it is for the wave to be detected as an extra contribution to the previously mentioned number of periods. A more careful and less conceptual discussion of the same nature will reveal that the signal strengths that should be compared with the sensitivities of LIGO are $h\sqrt{n}$ [26].

To calculate the characteristic signal strength of a gravitational wave, we must first find an estimate of the number n of periods a source spends around a certain frequency f. This estimate is given by the relation[27]

$$n = f^2 \left(\frac{df}{dt} \right)^{-1} \qquad \textbf{(6.2.5.13)}$$

But since the time derivative of the frequency satisfy the relation

$$\frac{df}{dt} = \frac{\partial f}{\partial T}\frac{\partial T}{\partial t} = -\frac{1}{T^2}\frac{dT}{dt} = -f^2 \frac{dT}{dt} \qquad \textbf{(6.2.5.14)}$$

the number n can be calculated by using the equations (6.2.5.6), (6.2.5.13) and (6.2.5.14) in the following way

$$n = -\left(\frac{dT}{dt} \right)^{-1} = \frac{T^{5/3}}{A} = \frac{5c^5 T^{5/3}}{192\pi (G2\pi)^{5/3} \mu (m_1 + m_2)^{2/3}} \qquad \textbf{(6.2.5.15)}$$

The expression (6.2.5.15) together with equation (6.2.5.3) then imply that

$$h_c = h\sqrt{n} = \frac{4}{\pi^{1/6} r}\sqrt{\frac{5}{192\pi}}\frac{G^{5/6}}{c^{3/2}}\frac{(m_1 m_2)^{1/2}}{(m_1+m_2)^{1/6}}\frac{1}{f^{1/6}} \qquad \textbf{(6.2.5.16)}$$

where f is the frequency of the gravitational wave. With the help of the above expression, we can now plot the characteristic signal strengths of the gravitational waves emitted by black hole binaries ($m_1 = m_2 = 10M_{Solar}$) and Neutron star binaries ($m_1 = m_2 = 1.4M_{Solar}$), as they spiral together and coalesce (see figure 14, below)

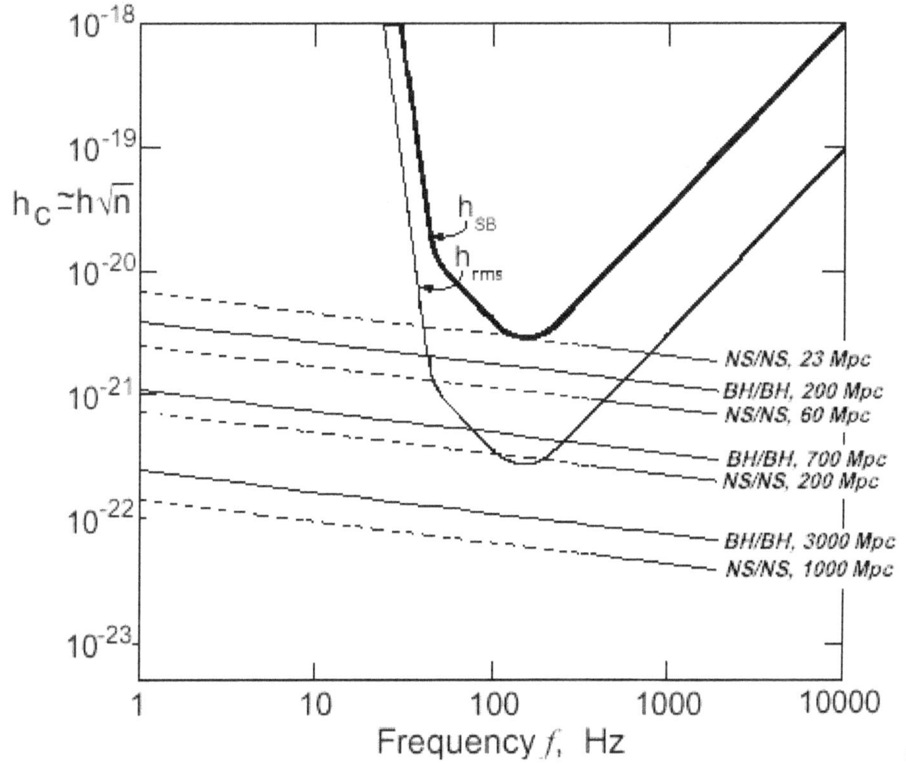

(figure 14)

With the help of the diagram in figure 14, we can compare the characteristic signal strengths of the coalescing binaries with the sensitivity curves of LIGO, and come to the conclusion that the coalescing neutron stars at distances 23Mpc, 60Mpc and 200Mpc together with the coalescing black holes at distances 200Mpc and 700Mpc will be detectable by LIGO.

It is interesting to compare the estimates of the characteristic signal strengths that have been made here, with more realistic estimates. Such estimates have been made by Kip S. Thorne [27], and these are shown in figure 15 on the next page.

The first difference between our estimates and the ones made by Thorne, is that we have assumed optimal orientation and location of the gravitational wave source, whereas Thorne has assumed an arbitrary orientation of the source. From this orientation, he has then calculated the mean value of the amplitudes for the waves travelling towards the detector, by averaging over all possible orientations[29].

The second and crucial difference is that Thorne has been using estimates of the interferometers' signal to noise ratios, in calculating the characteristic signal strengths[30], whereas we have only multiplied the amplitudes of the waves with the factor $\sqrt{n}$, in order to obtain the characteristic signal strengths. Despite these differences, the estimates made here and the ones made by Thorne (fig 15) predict almost the same characteristic signal strengths. The main reason for this is that both estimates are based on the quadrupole approximation (the weak field approximation).

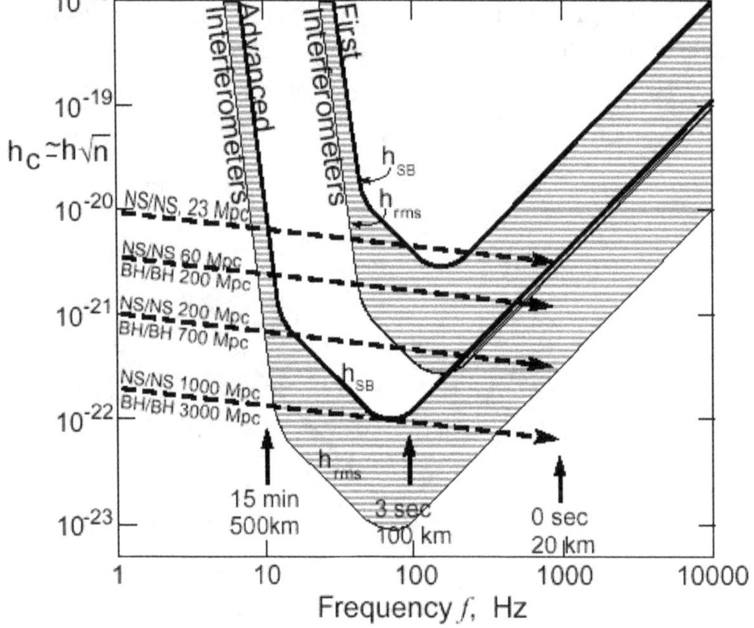

(figure 15)

The limits of the quadrupole approximation

Since the characteristic signal strengths exposed in figures 14 and 15 have been derived under the assumption that the gravitational field is weak, i.e. $h << 1$, it is of interest to us to explore the limits of this approximation. This can be done by using the relation (6.2.5.3) together with the weak field assumption $h << 1$, this reveals the condition

$$\frac{(16\pi)^{2/3} G^{5/3} \mu}{c^4 r} \left(\frac{m_1 + m_2}{T} \right)^{2/3} << 1 \qquad \textbf{(6.2.5.17)}$$

Furthermore by assuming that $r = a$ (where a is the distance between the stars / Black holes), and by using Kepler's third law to express the distance a in terms of the period T, we get the following weak field condition for the frequencies of the waves

$$f << \frac{c^3}{(32\pi^2)^{1/2} G \mu^{3/4} (m_1 + m_2)^{1/4}} \qquad \textbf{(6.2.5.18)}$$

With the help of this condition we can estimate that the weak field approximation used on coalescing black hole binaries ($m_1 = m_2 = 10 M_{Solar}$) is valid for frequencies $f << 1.6\,KHz$, and if it is used on coalescing neutron star binaries ($m_1 = m_2 = 1.4 M_{Solar}$) it is valid whenever $f << 11\,KHz$.

These rather rough estimates can be compared with less naïve estimates[31]. These reveal that the weak field approximation utilized on coalescing black hole and neutron star binaries is valid for frequencies $f << 1\,KHz$.

7 Summary

In contrary to electromagnetic radiation, gravitational radiation can travel through our universe without being easily scattered or absorbed by intervening matter. This means that the detection of gravitational radiation will enable us to study objects and phenomena that are invisible to the electromagnetic detectors. An example of such objects are the cores of supernovae, which are being electromagnetically disguised by dense layers of surrounding stellar gas.

Until September 2015, little evidence existed for the space time warpage of black holes and neutron stars, and since this warpage is so central to the study of general relativity, astronomers and theoretical physicists all over the world waited with great anticipation for the completion of the first gravitational wave observatories. It was expected that the first detectors would be operational and begin their first gravitational wave searches in 2002[32]. Furthermore, the second generation of LIGO interferometers (the advanced interferometers) were expected to be completed 2005-2006. It was estimated that these detectors would have sensitivities to gravitational waves improved by a factor ≈ 10 (see figure 15, page 62). All these expectations were fulfilled when, 11th February 2016, it was announced that the gravitational waves of two coalescing black holes of approximately 30 solar masses each had been detected on September 14, 2015 at 5:51 am Eastern Daylight time. A new era in physics and astronomy had begun.

Appendix A

In this appendix the identity

$$\left(D_{xx}-D_{yy}\right)^2+4\left(D_{xy}\right)^2=2D^{ik}D_{ik}-4D_z{}^iD_{iz}+\left(D_{zz}\right)^2 \quad \textbf{(A1)}$$

will be proven. This is done with the following straight forward deduction

$$2D^{ik}D_{ik}-4D_z{}^iD_{iz}+(D_{zz})^2=2\left[(D_{xx})^2+(D_{yy})^2+(D_{zz})^2+2\left((D_{zx})^2+(D_{zy})^2+(D_{xy})^2\right)\right]-$$

$$-4D_z{}^iD_{ik}+(D_{zz})^2=2\left[(D_{xx})^2+(D_{yy})^2\right]+4\left[(D_{zx})^2+(D_{zy})^2+(D_{zz})^2\right]+4\left(D_{xy}\right)^2-4D_z{}^iD_{ik}-$$

$$-(D_{zz})^2=2\left[(D_{xx})^2+(D_{yy})^2\right]+4D_z{}^iD_{ik}+4\left(D_{xy}\right)^2-4D_z{}^iD_{ik}-\left(D_{xx}+D_{yy}\right)^2=$$

$$=2\left[(D_{xx})^2+(D_{yy})^2\right]-\left(D_{xx}+D_{yy}\right)^2+4\left(D_{xy}\right)^2=\left(D_{xx}-D_{yy}\right)^2+4\left(D_{xy}\right)^2 \quad \square$$

Observe that in the above deduction the identity $D_{zz}=-\left(D_{xx}+D_{yy}\right)$ has been used.

Appendix B

In this appendix a proof will be presented for the following identities

$$\oiint_S n^k n^l dS=\frac{4\pi r^2}{3}\delta^{kl} \quad \textbf{(B1)} \quad ; \quad \oiint_S n^k n^l n^i n^j dS=\frac{4\pi r^2}{15}\left(\delta^{kl}\delta^{ij}+\delta^{ik}\delta^{jl}+\delta^{il}\delta^{jk}\right) \quad \textbf{(B2)}$$

where *S* is the surface of a sphere of radius *r*, and

$n^x=\sin(\theta)\cos(\varphi)$; $n^y=\sin(\theta)\sin(\varphi)$; $n^z=\cos(\theta)$ (i.e. the components of the unit normal to the sphere, expressed in spherical coordinates)

Proof of the identity (B1)

Due to spherical symmetry, the integral in the identity (B1) can only obtain two different values. One value corresponding to the case $k=l$, and the other value corresponding to the case $k\neq l$.

a) The case $k=l$

For the sake of simplicity we choose $k=l=z$, and then calculate the integral in (B1)

$$\oiint_S \left(n^z\right)^2 dS=r^2\int_0^{\pi}\int_0^{2\pi}\cos^2(\theta)\sin(\theta)d\theta d\phi=\frac{2\pi r^2}{3}\left[-\cos^3(\theta)\right]_0^{\pi}=\frac{4\pi r^2}{3} \quad \textbf{(B3)}$$

b) The case $k \neq l$

Let us choose $k = x$ and $l = z$, if we then calculate the integral in (B1) we get

$$\oint\!\!\oint_S n^x n^z dS = r^2 \int_0^{\pi}\int_0^{2\pi} \sin^2(\theta)\cos(\theta)\cos(\varphi)d\theta d\varphi = \frac{r^2}{3}\left[\sin^3(\theta)\right]_0^{\pi}\left[\sin(\varphi)\right]_0^{2\pi} = 0 \qquad \textbf{(B4)}$$

Finally, the previously mentioned symmetry together with (B3) and (B4) must imply that

$$\oint\!\!\oint_S n^k n^l dS = \frac{4\pi r^2}{3}\delta^{kl} \qquad \square$$

Proof of the identity (B2)

Due to spherical symmetry the integral (B2) can only be expected to obtain four different values. These values correspond to the cases when

i) Two indexes are equal
ii) Three indexes are equal
iii) There are two distinct pairs of indexes, where the indexes in each pair are equal
iv) All four indexes are equal

The case (i):

Let us choose $k = l = x$, $i = y$ and $j = z$, then the integral in (B2) becomes

$$\oint\!\!\oint_S \left(n^x\right)^2 n^y n^z dS = r^2 \int_0^{\pi}\int_0^{2\pi} \sin^4(\theta)\cos(\theta)\cos^2(\phi)\sin(\phi)d\theta d\phi = \frac{r^2}{15}\left[\sin^5(\theta)\right]_0^{\pi}\left[-\cos^3(\phi)\right]_0^{2\pi} = 0 \qquad \textbf{(B5)}$$

The case (ii):

Let us choose $k = l = i = z$ and $j = x$, then if we calculate the integral in (B2) we get

$$\oint\!\!\oint_S \left(n^z\right)^3 n^x dS = r^2 \int_0^{\pi}\int_0^{2\pi} \cos^3(\theta)\sin^2(\theta)\cos(\phi)d\theta d\phi = r^2\left[\sin(\phi)\right]_0^{2\pi}\int_0^{\pi}\cos^3(\theta)\sin^2(\theta)d\theta = 0 \qquad \textbf{(B6)}$$

The case (iii):

Let us choose $k = l = x$ and $i = j = z$, then the integral in (B2) becomes

$$\oiint_S (n^x)^2 (n^z)^2 \, dS = r^2 \int_0^{\pi} \int_0^{2\pi} \sin^3(\theta)\cos^2(\theta)\cos^2(\phi) d\theta d\phi =$$

(B7)

$$= r^2 \left[\frac{\phi}{2} + \frac{\sin(2\phi)}{4} \right]_0^{2\pi} \left[-\frac{\cos^3(\theta)}{3} + \frac{\cos^5(\theta)}{5} \right]_0^{\pi} = \frac{4\pi r^2}{15}$$

The case (iv):

For the sake of simplicity we choose $k = l = i = j = z$, then the integral in (B2) becomes

$$\oiint_S (n^z)^4 \, dS = r^2 \int_0^{\pi} \int_0^{2\pi} \cos^4(\theta)\sin(\theta) d\theta d\phi = \frac{2\pi r^2}{5} \left[-\cos^5(\theta) \right]_0^{\pi} = \frac{4\pi r^2}{5} \qquad \textbf{(B8)}$$

Finally, the previously mentioned symmetry together with (B5)-(B8) must imply that

$$\oiint_S n^k n^l n^i n^j \, dS = \frac{4\pi r^2}{15} \left(\delta^{kl}\delta^{ij} + \delta^{ik}\delta^{jl} + \delta^{il}\delta^{jk} \right) \quad \square$$

References

[1] M. Lipschutz, Theory and problems of differential geometry, page 213, Schaum's outline series, McGraw-Hill, 1969.
[2] M. Lipschutz, Theory and problems of differential geometry, page 201-203, Schaum's outline series, McGraw-Hill, 1969.
[3] B.F Schutz, A first course in general relativity, page 158-160, Cambridge university press, 1985.
[4] B.F Schutz, A first course in general relativity, page 160-162, Cambridge university press, 1985.
[5] B.F Schutz, A first course in general relativity, page 172-173, Cambridge university press, 1985.
[6] B.F Schutz, A first course in general relativity, page 110, Cambridge university press, 1985.

[7] B.F Schutz, A first course in general relativity, page 104-105, Cambridge university press, 1985.
[8] R. M Wald, General relativity, page 71-73, university of Chicago press, 1984.
[9] B.F Schutz, A first course in general relativity, page 205-206, Cambridge university press, 1985.
[10] B.F Schutz, A first course in general relativity, page 205, Cambridge university press, 1985.
[11] S.M Caroll, Lecture notes in general relativity, chapter 6 page 13, Institute for theoretical physics University of California Santa Barbara, http://itp.ucsb.edu/ caroll/notes
[12] R.K Wangsness, Electromagnetic fields, 2:nd edition, page 469-471,Wiley & sons, 1986.
[13] B.F Schutz, A first course in general relativity, page 234-240, Cambridge university press, 1985.
[14] B.F Schutz, A first course in general relativity, page 225, Cambridge university press, 1985.
[15] G.H Sanders & D Beckett, Sky & Telescope, page 48, October 2000.
[16] P.S. Shawhan, The Search for Gravitational Waves with LIGO: Status and Plans – An Article for the Proceedings of the DPF2000 Meeting of the American Physical Society, page 1, document No P000021-00-E, Submitted to International Journal of Modern Physics A,11/03/2000.
[17] P.S. Shawhan, The Search for Gravitational Waves with LIGO: Status and Plans – An Article for the Proceedings of the DPF2000 Meeting of the American Physical Society, page 2, document No P000021-00-E, Submitted to International Journal of Modern Physics A,11/03/2000.
[19] K.S. Thorne, "Gravitational waves from compact bodies," in *Compact Stars in Binaries*, Proceedings of IAU Symposium 165, The Hague, Netherlands, August 1994, edited by J. van Paradijs, E. van den Heuvel, and E. Kuulkers (Dordrecht, Boston, 1996), page 157.

[20] K.S. Thorne, "Gravitational waves," in *Proceedings of the Snowmass'94 Summer Study on Particle and Nuclear Astrophysics and Cosmology*, eds. E. W. Kolb and R. Peccei (World Scientific, Singapore, 1995), pp. 404-405; also published in *Particle Physics, Astrophysics & Cosmology*, Proceedings of the SLAC Summer Institute on Particle Physics, eds. Jennifer Chan & Lilian DePorcel (SLAC-Report-484, Stanford Linear Accelerator Center, Stanford, California, 1996).

[21] K.S. Thorne, "Gravitational waves," in *Proceedings of the Snowmass'94 Summer Study on Particle and Nuclear Astrophysics and Cosmology*, eds. E. W. Kolb and R. Peccei (World Scientific, Singapore, 1995), page 405-406; also published in *Particle Physics, Astrophysics & Cosmology*, Proceedings of the SLAC Summer Institute on Particle Physics, eds. Jennifer Chan & Lilian DePorcel (SLAC-Report-484, Stanford Linear Accelerator Center, Stanford, California, 1996).

[22] K.S. Thorne, "Gravitational waves," in *Proceedings of the Snowmass'94 Summer Study on Particle and Nuclear Astrophysics and Cosmology*, eds. E. W. Kolb and R. Peccei (World Scientific, Singapore, 1995), page 406; also published in *Particle Physics, Astrophysics & Cosmology*, Proceedings of the SLAC Summer Institute on Particle Physics, eds. Jennifer Chan & Lilian DePorcel (SLAC-Report-484, Stanford Linear Accelerator Center, Stanford, California, 1996).

[23] C. Brif, Organization of Thermal Noise in Multi-Loop Pendulum Suspensions for Use in Interferometric Gravitational-Wave Detectors, Laser Interferometer Gravitational Wave Observatory, document No P990028-00-D, 07/29/1999.

[24] K.S. Thorne, "Gravitational waves," in *Proceedings of the Snowmass'94 Summer Study on Particle and Nuclear Astrophysics and Cosmology*, eds. E. W. Kolb and R. Peccei (World Scientific, Singapore, 1995), page. 408; also published in *Particle Physics, Astrophysics & Cosmology*, Proceedings of the SLAC Summer Institute on Particle Physics, eds. Jennifer Chan & Lilian DePorcel (SLAC-Report-484, Stanford Linear Accelerator Center, Stanford, California, 1996).

[25] K.S. Thorne, "Gravitational waves from compact bodies," in *Compact Stars in Binaries*, Proceedings of IAU Symposium 165, The Hague, Netherlands, August 1994, edited by J. van Paradijs, E. van den Heuvel, and E. Kuulkers (Dordrecht, Boston, 1996), page 159.

[26] K.S Thorne. In S.W. Hawking and W. Israel, editors, Three Hundred Years of Gravitation, page 381. Cambridge University press, 1987.

[27] K.S. Thorne, "Gravitational waves from compact bodies," in *Compact Stars in Binaries*, Proceedings of IAU Symposium 165, The Hague, Netherlands, August 1994, edited by J. van Paradijs, E. van den Heuvel, and E. Kuulkers (Dordrecht, Boston, 1996), page 163.

[28] K.S. Thorne, "Gravitational waves from compact bodies," in *Compact Stars in Binaries*, Proceedings of IAU Symposium 165, The Hague, Netherlands, August 1994, edited by J. van Paradijs, E. van den Heuvel, and E. Kuulkers (Dordrecht, Boston, 1996), page. 160.

[29] K.S Thorne. In S.W. Hawking and W. Israel, editors, Three Hundred Years of Gravitation, page 380. Cambridge University press, 1987.

[30] K.S Thorne. In S.W. Hawking and W. Israel, editors, Three Hundred Years of Gravitation, pages 366-381. Cambridge University press, 1987.

[31] K.S Thorne. In S.W. Hawking and W. Israel, editors, Three Hundred Years of Gravitation, page 379. Cambridge University press, 1987.

[32] G.H Sanders & D Beckett, Sky & Telescope, page 41-48, October 2000.

www.ingramcontent.com/pod-product-compliance
Lightning Source LLC
Chambersburg PA
CBHW081836170526
45167CB00007B/2831
* 9 7 8 1 5 4 2 7 5 0 2 2 6 *